The Environment

a dictionary of the
world around us

Arrow Reference Series

General Editor: Chris Cook

The Environment

a dictionary of the world around us

Geoffrey Holister
and
Andrew Porteous

Arrow Books

Arrow Books Limited
3 Fitzroy Square, London W1

An imprint of the Hutchinson Publishing Group

London Melbourne Sydney Auckland
Wellington Johannesburg and agencies
throughout the world

First published 1976
© Andrew Porteous and Geoffrey Holister 1976

Set in Monotype Times

Made and printed in Great Britain
by The Anchor Press Ltd,
Tiptree, Essex

ISBN 0 09 913470 5

'May you live in interesting times!'
OLD CHINESE CURSE

Acknowledgements

We would like to thank those many people who have helped us in the compilation of this dictionary. In particular we would like to thank Judy Anderson for her research and J. Roger Tagg for his helpful comments on the manuscript.

We are also grateful to Professor Michael Hussey and Professor William Gosling for their advice and help; and Sue Hogg for her invaluable assistance in making a coherent whole of our individual efforts.

Introduction

Just over 120 years ago, in the Northwest Territories of the United States, the Suquamish Indian tribe faced near extinction in its confrontation with the westward movement of white society. The chief of the tribe was a patriarch named Seattle. In his final negotiation with government agents forcing the Suquamish onto a reservation, Seattle said some notable things that lay ignored, as curious cultural irrelevancies, for over a hundred years. He said:

. . . If we do not sell, the white man may come with guns and take our land. How can you buy or sell the sky, the warmth of the land? The idea is strange to us. If we do not own the freshness of the air and the sparkle of the water, how can you buy them?

. . . If we sell you land, you must remember that it is sacred, and you must teach your children that it is sacred, and that each ghostly reflection in the clear water of the lakes tells of events and memories in the life of my people.

. . . If we sell you our land, you must remember and tell your children, that the rivers are our brothers, and yours, and you must henceforth give the rivers the kindness you would give any brother.

. . . Teach your children what we have taught our children, that the earth is our mother. Whatever befalls the earth, befalls the sons of the earth. If men spit upon the earth, they spit upon themselves.

. . . The earth does not belong to man; man belongs to the earth . . . all things are connected. . . . Man did not weave the web of life; he is merely a strand in it. Whatever he does to the web, he does to himself.

This book is written in the hope that it may contribute, however minutely, to a rediscovery of the deep insights exemplified by Chief Seattle's words.

If, in reading this work, you conclude that its authors are pessimists – the current phrase in an era of environmental backlash is 'doommongers' – then you will be misinterpreting our position, which is one of profound concern tinged with a touch – but just a touch – of optimism.

The present attitude of governments and their specialist advisers does not, at first sight, give cause for any optimism at all. One is reminded of the man who fell from the top of the Empire State Building and, as he passed a window on the fiftieth floor, was heard to mutter 'So far – so good.' But there are hopeful signs. In higher education there is a move towards multidisciplinary studies and away from increasing specialization. The emergence of intellectually influential groups such as Friends of the Earth, the attempts by Dr E. F. Schumacher and his colleagues to develop an 'intermediate technology', even the painful pangs of growing understanding brought about by the current oil crisis – all these things give us some cause for hope.

In writing this book, we have had to select information from an almost infinite source of data, much of it conflicting. In making this selection we have naturally had to apply our own value judgements as to which data are relevant and important and which are not. In doing this we have attempted to be as objective as possible, but under such circumstances perfect objectivity is quite impossible. We cannot assess the value of data without passing a tacit judgement on the character of the source. Objectivity does not consist in giving equal weight to all statements, and often entails taking sides against those whom one judges to be deluding themselves or resorting to deception.

We do not attempt in this book to give solutions to the problems that we must face. Indeed it is doubtful whether a 'correct' solution exists – one man's solution is another man's problem in such a complex world. Our aim is to attempt to demonstrate the nature of the issues. As our understanding of these becomes clearer, we at least begin to appreciate a number of things that we must obviously *not* do, and that is a start of sorts.

Environmental problems are essentially multi-faceted and demand at least a nodding acquaintance with many previously separate specialisms – ecology, economics, sociology, technology, physics, chemistry, and so on. The world is an enormously complex system and it is in the nature of complex systems that the characteristics of the connections between the constituent parts are often more important than the nature of the separate parts themselves.

This book is designed for a multi-access approach on the part of the reader – it can be dipped into, as well as read straight through. The reason for this format is the obvious diversity of backgrounds and interests of the readers. Most of you who read this book will have some specialist knowledge of some aspect of our technological society, and are likely to be interested in one particular aspect of our environment more than another. The format of this book allows you to select those areas of interest, although because of the complex nature of our social, economic, political and ecological background, you will find that *wherever* you start in this book, you will be led inexorably by the references to other areas of the problem that are probably new to you. But that is the nature of the environment; everything is related, in some way or another, to everything else.

As chief Seattle said: '. . . all things are connected . . .'

Geoffrey Holister
Andrew Porteous

The Open University, 1976

Abbreviations

ADI	Acceptable daily intake
A/G ratio	Arithmetic–geometric ratio
AN	*Aspergillus niger*
BHC	Benzene hexachloride
BOD	Biochemical oxygen demand
Ci	curie
CNL	Corrected noise level
COD	Chemical oxygen demand
c.o.p.	Coefficient(s) of performance
dB	decibel
dB(A)	decibels A-scale
DDE	1, 1-di-chloro-2. 2-bis (p-chlorophenyl) ethylene
DDT	Dichlorodiphenyltrichloroethane
DDVP	Dichlorvos-2, 2-dichlorovinyl dimethyl phosphate
DES	Diethylstilboestrol
DNA	Deoxyribonucleic acid
DO	Dissolved oxygen
EDTA	Ethylenediamine-tetracetate
E–M pathway	Embden–Meyerhof pathway
EPA	Environmental Protection Agency (USA)
EPNdB	Effective perceived noise level
ERTS	Earth Resources Technology Satellite
FMC	Field moisture capacity
GNP	Gross national product
Hz	hertz
ICRP	International Commission for Radiological Protection
IUD	Intrauterine device

IWC	International Whaling Commission
L_{10}	10 per cent level (road traffic noise index)
LC_{50}	Lethal concentration (50 per cent survival)
LD_{50}	Lethal dose (50 per cent survival)
MAC	Maximum allowable concentration
MHD	Magnetohydrodynamic generator
NNI	Noise and number index
OECD	Organization for Economic Cooperation and Development
PAN	Peroxyacetylnitrate
PCB	Polychlorinated biphenyl(s)
PNdB	Perceived noise decibels
pphm	parts per hundred million
ppm	parts per million
PTFE	Teflon
PVC	Polyvinyl chloride
RBE	Relative biological effectiveness
R/P ratio	Reserves–production ratio
SCP	Single cell protein
SI	International System of Units
SMD	Soil moisture deficit
STP	Standard temperature and pressure
TCDD	Tetrachlorodibenzo-p-dioxin
TLV	Threshold limiting value
TNI	Traffic noise index
UDC	Underdeveloped country
VCM	Vinyl chloride monomer
WHO	World Health Organization
WLM	Working level month

Abbreviated Table of Elements

Aluminium	**Al**	Helium	**He**	Polonium	**Po**
Antimony	**Sb**	Hydrogen	**H**	Radium	**Ra**
Argon	**Ar**	Iodine	**I**	Radon	**Rn**
Arsenic	**As**	Iron	**Fe**	Rubidium	**Rb**
Barium	**Ba**	Krypton	**Kr**	Ruthenium	**Ru**
Beryllium	**Be**	Lead	**Pb**	Selenium	**Se**
Bromine	**Br**	Magnesium	**Mg**	Silicon	**Si**
Cadmium	**Cd**	Manganese	**Mn**	Silver	**Ag**
Caesium	**Cs**	Mercury	**Hg**	Sodium	**Na**
Calcium	**Ca**	Molybdenum	**Mo**	Strontium	**Sr**
Carbon	**C**	Neon	**Ne**	Sulphur	**S**
Chlorine	**Cl**	Nickel	**Ni**	Tin	**Sn**
Chromium	**Cr**	Nitrogen	**N**	Uranium	**U**
Cobalt	**Co**	Oxygen	**O**	Vanadium	**V**
Copper	**Cu**	Phosphorous	**P**	Xenon	**Xe**
Fluorine	**F**	Plutonium	**Pu**	Zinc	**Zn**
Gold	**Au**	Potassium	**K**		

Table of Prefixes for SI Units

Prefix	Symbol	Factor
tera	T	$10^{12} = 1\ 000\ 000\ 000\ 000$
giga	G	$10^9 = 1\ 000\ 000\ 000$
mega	M	$10^6 = 1\ 000\ 000$
kilo	k	$10^3 = 1000$
hecto	h	$10^2 = 100$
deca	da	$10^1 = 10$
deci	d	$10^{-1} = 0 \cdot 1$
centi	c	$10^{-2} = 0 \cdot 01$
milli	m	$10^{-3} = 0 \cdot 001$
micro	μ	$10^{-6} = 0 \cdot 000\ 001$
nano	n	$10^{-9} = 0 \cdot 000\ 000\ 001$
pico	p	$10^{-12} = 0 \cdot 000\ 000\ 000\ 001$
femto	f	$10^{-15} = 0 \cdot 000\ 000\ 000\ 000\ 001$
atto	a	$10^{-18} = 0 \cdot 000\ 000\ 000\ 000\ 000\ 001$

Conversion Table for SI and British Units

Physical property	British unit	SI unit	SI unit	British unit	Cgs unit
length	1 ft	0·305 m	1 m	3·28 ft	100 cm
	1 mile	1·61 km	1 km	0·621 mile	10^5 cm
area	1 ft^2	0·0929 m^2	1 m^2	10·76 ft^2	10^4 cm^2
	1 acre	4047·0 m^2	1 km^2	$2·471 \times 10^2$ acres.	
	1 acre	0·405 ha	1 ha	2·471 acres	
volume	1 ft^3	0·0283 m^3	1 m^3	35·31 ft^3	10^6 cm^3
	1 gall (UK)	4·55 litres	1 litre	0·220 gallon (UK)	10^3 cm^3
mass	1 lb	0·454 kg	1 kg	2·204 lb	10^3 g
density	1 lb/ft^3	16·02 kgm^{-3}	1 kg m^{-3}	0·0624 lb/ft^3	10^3 g cm^{-3}
time	1 sec	1 s	1 s — a defined fraction of a solar day 3600 s = 1 h	1 second	1s
velocity	1 ft/s	0·305 m s^{-1}	1 m s^{-1}	3·28 ft/s	100 cm/s
			1 km h^{-1}	0·911 ft/s	27·8 cm/s
	1 mile/h	1·61 km h^{-1}	1 km h^{-1}	0·621 mile/h	
force (mass × acceleration)	1 lbf	4·45 N	1 N(kgm s^{-2})	0·225 lbf	10^5 dyn (dynes) (1 dyn = gcm s^{-2})
pressure (force per unit area)	1 lbf/ft^2	47·5 N m^{-2}	1 N m^2 (1 pascal – Pa)	0·0207 lbf/ft^2	10 dyn cm^{-2}
pressure in meteorology and acoustics			1 bar	29·53 ins Hg	750 cm Hg
			10^5 N m^{-2}		(10^6 dyn cm^{-2})
			1·013 bar	29·91 ins Hg	76·0 cm Hg
work (force × distance) energy	1 ft lbf	1·352 J	1 J (Nm) (joule)	0·738 ft lbf	10^7 ergs
heat equivalent	1 Btu	1·055 kJ	1 kJ	0·948 Btu	10^{10} ergs
power (rate of doing work)	1 ft lbf/s	1·356 W	1W(J s^{-1})	0·738 ft lbf/s	10^7 ergs s^{-1}
	1 hp	0·746 kW	1 kW	1·34 hp	10^{10} erg s^{-1}
	1 million gall/d	0·0526 m^3 s^{-1} (Cumecs)	1×10^6 m^{-3} day^{-1} (Cumecs)	19·01 million gall/d	10^6 cm^3 s^{-1}
				220 mgd	
flow rate	1 gall/min	0·0761 s^{-1}	11 s^{-1}	13·20 gall/min	10^3 cm^3 s^{-1}

A

Abortion. Termination of a pregnancy in its early stages. The commonest methods employed are curettage (scraping) of the uterus, or the use of a vacuum device for removing the foetus.

Although modern methods of birth control are relatively effective (▷CONTRACEPTION), it is nevertheless arguably the case that abortion is the commonest form of birth control world wide, even in countries where contraceptive devices are readily available.

In spite of the tenacity of ingrained religious and ideological beliefs and a natural repugnance towards the artificial termination of a potential human life, there is little doubt that in an overpopulated world the best solution for ensuring world peace is the widespread promotion of contraception and abortion. (▷BIRTH CONTROL; POPULATION GROWTH.)

Absolute temperature scale. ▷TEMPERATURE.

Absorption (1). The taking up, usually, of a liquid or gas into the body of another material (the absorbent). Thus, for instance, an air pollutant may be removed by absorption in a suitable solvent.

Absorption (2). The taking up of radiation by a material it encounters or passes through. This can be the basis of measuring the mass of a substance or identifying a number of substances.

Acceptable daily intake (ADI). The acceptable daily intake of a chemical is the daily intake which, during an entire lifetime, appears to be without risk on the basis of all known facts at the time. It is expressed in milligrams of chemical per kilogram of body weight ((mg kg^{-1}). ADIs may be unconditional, conditional, or temporary.

Acid dewpoint. As commonly used, this term applies to the temperature at which dilute acid (e.g. sulphuric acid) appears as a condensate (liquid droplets) when a flue gas containing sulphur trioxide and water vapour is cooled below saturation. This is related to the moisture and sulphur trioxide content of the flue gas. Therefore, gases must be released to atmosphere well above this temperature otherwise substantial corrosion will take place in the stack. The corrosion of many incinerator installations is due to the flue gases being cooled below the acid dewpoint.

Acid mine drainage. Many mining operations, particularly those that work sulphide ores, e.g. nickel, copper or coal mining where pyrites (iron sulphide) are present, can, through a combination of air and moisture, form acidic and metal-bearing solutions. This combination of acids and metals can have severe local effects on the ecology of streams and rivers, and the metals can enter food chains and further affect life. If, as is likely, iron sulphates are formed, this will result in brown coating on rocks and stream beds, further despoiling the appearance of the area.

In the UK the problems of acid mine drainage are not severe, but in nickel and copper mining areas (e.g. Sudbury, Ontario) the devastation has to be seen to be believed.

The only way of controlling or preventing acid mine drainage other than not working the ores is to control the pH ($\Diamond$pH) by raising it with lime and neutralizing the drainage water. The tips can then be grassed and the subsequent exclusion of air will prevent the oxidizing of the sulphides. The successful application of revegetation techniques requires the initial neutralization of the acidic effluents.

Acid rain. Most rainfall is slightly acidic due to carbonic acid from the carbon dioxide content of the atmosphere, but 'acid rain' in the pollution sense is rainfall contaminated usually with sulphuric acid due to the sulphur oxides emitted from the sulphur content of fuels and industrial emissions. The effects of this rainfall can be serious on crops, as it can leach ($\Diamond$LEACHING) calcium ions from both leaves and soil. Also the sulphur dioxide can thereby enter the cell walls of plants and upset the ionic balance. Conifer growth near industrial installations that cause acid rainfall can sometimes be half the normal growth rate.

For economic damage in the UK, for example, see *Final Report of the Beaver Committee*, Cmd 9322, HMSO, 1954.
See also 'Recognition of Air Pollution Injury to Vegetation: A Pictorial Atlas', Information Report No. 1, TR7, Agricultural Committee, US Air Pollution Control Association, 1970.

Actinides. This group of elements, with ATOMIC NUMBERS from 89 to 103, includes PLUTONIUM. All members of this group are produced in nuclear reactions or by radioactive decay.

Activated carbon. Carbon obtained from vegetable or animal matter by roasting in a vacuum furnace. Its porous nature gives it a very high surface area per unit mass – as much as 1000 square metres per gramme, which is 10 million times the surface area of one gramme of water in an open container. For this reason, the substance is a very good adsorbent for aromatic and unsaturated aliphatic compounds. It is extensively used for odour control and air-freshening applications, and can adsorb large quantities of gases.

Activated sludge. The sludge removed from the activated sludge sewage treatment process ($\Diamond$ SEWAGE EFFLUENT TREATMENT). It consists of BACTERIA and PROTOZOA which can live and multiply on the sewage. Because of this multiplication, the excess organisms require continuous removal. Part of the sludge is returned to the raw sewage (hence the term activated sludge), and part (approximately 90 per cent) is sent for disposal.

Activated-sludge process. $\Diamond$ SEWAGE EFFLUENT TREATMENT.

Additives. $\Diamond$ FOOD ADDITIVES.

Adreno-cortical stress. $\Diamond$ CROWDING.

Adsorption. A phenomenon in which molecules of a substance (the adsorbate) are taken up and held on the surface of a material (the adsorbent) (compare ABSORPTION). ($\Diamond$ ACTIVATED CARBON.)

Advertising. Any means of informing the general public, persuading them to act in a prescribed manner, or obtaining public favour or notoriety. Advertising is a principal weapon in the armoury of business interests intent on the management of consumer demand in that section of our industry which is concerned with the production of non-essential goods. The advertising industry hotly deny this interpretation of their role and insist that the dissemination of useful information is still one of their major functions. However, as Galbraith has observed, 'only a gravely retarded citizen can need to be told that the American Tobacco Company has cigarettes for sale.'

The dependence of Western economies upon this production of non-essential goods means that goods that merely satisfy basic needs (food, clothing, shelter) comprise a small and diminishing part of all production. Thus, the earth's limited resources are being converted into profitable rubbish at an accelerating rate, yet the active suppression of demand would undoubtedly cause grave economic hardship, so dependent is our economy on the production of that which can be profitably sold.

For anyone interested in the role of advertising in economic life, J. K. Galbraith's *The Affluent Society*, Hamish Hamilton, 1958; Penguin Books, 1970, is required reading.

Aerobic process. Metabolism in the presence of air or oxygen. For example, water in an aerobic stream contains DISSOLVED OXYGEN and therefore organisms utilizing this can oxidize organic wastes to simple compounds as below:

$$\text{organic materials} + O_2 \xrightarrow{\text{micro-organisms}} CO_2 + H_2O + NH_3 + \text{micro-organisms.}$$

Ammonia (NH_3) can then be further oxidized to a compound containing the nitrate (NO_3) by bacteria and the sewage is thereby totally decomposed.

With ANAEROBIC PROCESSES a totally different decay mechanism occurs. FERMENTATION can be aerobic or anaerobic.

Aerosols. Minute particles in suspension, usually as a mist suspended in air or other gases. They are less than one micron in size.

Aerosol propellant. Colloquially, the pressurized system used to disperse hair spray, deodorant, etc. More accurately, an inert liquid with a low boiling point, from the CHLOROFLUOROMETHANES or hydrocarbons, which vaporizes instantaneously at room temperatures on release of pressure. When the pressure in the aerosol canister is released, the vapour carries the aerosol (or cloud of moisture particles) of the desired substance to its target. The propellant then disperses into the atmosphere, and compounds containing chlorine have been cited as a potential hazard to the earth's OZONE SHIELD because of the free CHLORINE liberated in the upper atmosphere as a result of ultra-violet radiation.

Oregon State is to ban the use of fluorocarbon propellants from February 1977; other states are expected to do the same. A recent statement from the UK government acknowledges that they may pose a threat.

Other propellants are available, such as butane, which is highly inflammable and therefore only suitable for spraying water-based products. Carbon dioxide, nitrous oxide and nitrogen are possibly suitable for most spray purposes and are inert, but there is no commercial incentive for their adoption, and as yet no legal requirement that this be done to protect the environment.

Age-specific birth and death rates. In any consideration of a country's population and its potential growth there are a number of critical factors: size, growth rate, age structure, sex ratio and distribution. Birth and death rates expressed simply in terms of births or deaths per thousand population are known technically as crude birth and death rates and, because they do not take into account the age structure of the population, can be misleading. Age-specific birth and death rates are simply rates applicable to a specified age class.

It is the age structure of a population which is probably the most critical factor in any attempt to predict future population growth. Figure 1 compares the age structure of the population of Mauritius and the United Kingdom in 1959. They are fairly characteristic of the age structures of an underdeveloped country and a developed country respectively. Because the profiles are based on proportions, they both

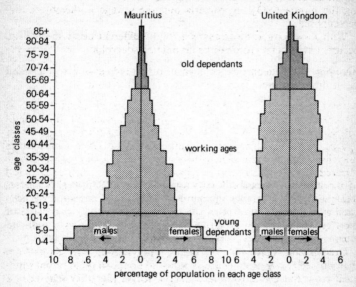

Figure 1. Age structure of the population of Mauritius and of the United Kingdom in 1959. (After *Population Bulletin*, vol. 18, no. 51.)

have the same area. Note the very large percentage of children in the Mauritius profile. It is here that the potential for future explosive population growth lies. Beware, therefore, of arguments based in current data for crude birth rates – a drop in current figures for crude birth rates does not, by itself, mean an end to population growth.

Aggression. ⟡CROWDING.

Agricultural development. It is frequently argued that projected population growth could be catered for by an increase in world food output through properly organized agricultural development. Such development, it is suggested, could be directed towards increasing yields from existing arable land (◊GREEN REVOLUTION); bringing more new land under cultivation (◊ARABLE LAND, POTENTIAL OF); and harvesting food from the sea (◊FOOD FROM THE SEA).

Such arguments would have more weight if there were signs of any serious attempt to alleviate the hunger of the present 2400 million people who are either undernourished – that is, not receiving sufficient calories per day – or malnourished – that is, seriously lacking in one or more essential nutrients, most commonly protein.

Agricultural economics. The application of traditional economic criteria to the production of food. It is unfortunate that government policies are usually designed by specialists whose experiences are essentially urban and industrial, and who have little or no understanding of the complex ecological balances required for a stable agricultural system.

Agriculture cannot be regarded simply as an industry. The majority of our industries are concerned with the conversion of raw material – which they regard as virtually inexhaustible – into goods. The raw material of agriculture is the soil, which should be husbanded – not mined. (◊MODERN FARMING METHODS; ◊◊SOIL, FERTILITY AND EROSION OF.)

Agriculture, Energy and efficiency aspects. The effectiveness of modern agriculture is invariably measured in traditional economic terms – percentage return on invested capital, yields per acre, yields per man-year, etc. (◊AGRICULTURAL ECONOMICS). It is in these terms that claims for the high efficiency of modern industrialized agricultural techniques are made. At first sight, these claims are persuasive: in Britain, average wheat yields are up from 19·1 cwt per acre in 1946 to 28·2 cwt in 1968–9; barley up from 17·8 cwt per acre in 1946 to 27·4 cwt per acre in 1968–9; and with similar increases in most other field crops.

Unfortunately, closer investigation shows that when such figures are apportioned and averaged per agricultural worker they misrepresent the true facts because of the nature of the oversimplifications that are made. For instance, figures showing increased yields per farm worker ignore all of those workers who have moved from farms into agriculture-related industries, and are engaged in the production of the machines, fuel and chemicals which give the remaining farm workers their spurious efficiency.

A more realistic measure of agricultural efficiency is given by the relative inputs and outputs of energy to the total system.

This calculation is difficult when applied to a modern industrial-type agricultural system; nevertheless, the total energy budgets for British agriculture have been calculated (see Leach and Slesser, 1973) and compared with the efficiency of other agricultural systems (see Leach, 1974). Their general conclusions were that, where a fossil-fuel subsidy does not exist, human energy can be used very efficiently. For example, maize growers in Yucatan produce 13 to 29 units of food energy for each unit of (predominantly human) energy expended; primitive gardeners in Tsembaga, New Guinea, produce 20 units of energy for each unit expended. British wheat growers, on the other hand, produce 2·2 energy units for each unit expended; potato growers gain 1·1 units per unit expended; sugar beet, by the time it has been refined to white sugar, represents 0·49 units gained per unit expended; a battery egg, including the food value of the hen carcase at the end of her laying life, represents 0·16 units for each unit expended; and a broiler chicken, 0·11 units gained.

In terms of numbers fed per worker, it transpires that, because the number of people engaged in agriculture-related industries in Britain is approximately the same as the agricultural workforce of a peasant economy, our much vaunted agricultural efficiency is only about four times that of the Kalahari bushmen, who are hunters and gatherers (although, of course, such comparisons ignore differences in standards of living).

G. Leach and M. Slesser, *Energy Equivalent of Network Inputs to Food Producing Processors*, Strathclyde University, 1973.
G. Leach, *The Man–Food Equation*, Academic Press, 1974.

Airborne particulate matter. ⋄DUST.

Air classifier. Dry separation device used to separate shredded domestic refuse by density difference methods. It can be used in conjunction with ferrous-magnet and revolving screens to sort refuse for recycling. (⋄RECYCLING, Figure 53.)

Aircraft noise. Aircraft noise consists of a build-up to a peak level and then a fall-off, occurring at intervals, as opposed to the continuous but fluctuating noise from heavy road traffic. In social surveys around Heathrow Airport the annoyance caused by noise from air traffic was found to depend on the peak perceived noise levels and on the number of aircraft heard in a given period. Hence an index was derived termed NOISE AND NUMBER INDEX (NNI) which combines the two quantities according to a formula for a given period during the day. Some values of NNI and an indication of the conditions associated with them are listed below.

NNI   *Typical air-traffic conditions*

60   This occurs only close to airports where there are many over-flights at low altitude. Noise levels can interfere with sleep and conversation in ordinary houses and may also interfere even within sound insulated houses.

45   This occurs mostly near busy routes from airports. Many aircraft are heard at noise levels which can interfere with conversation in ordinary houses.

35   The overflying is typically irregular at noise levels which are noticeable and occasionally will be intrusive within ordinary houses.

($\lozenge$NOISE; SOUND; NOISE INDICES; HEARING; ROAD-TRAFFIC NOISE; INDUSTRIAL NOISE MEASUREMENT.)

Albedo.   The ratio of light reflected from a particle, planet or satellite to that falling on it. Therefore, it always has a value less than or equal to 1. It is also used in nuclear physics.

The albedo of the earth plays an important part in the earth's radiation balance and influences the MEAN ANNUAL TEMPERATURE, and therefore the CLIMATE, on both a local and global scale.

Aldehydes.   Organic compounds containing the group –CHO attached to a hydrocarbon. As air pollutants a number of them have an unpleasant smell, e.g. in diesel exhaust, and can be an irritant to nose and eyes; many can be poisonous. ($\lozenge$AUTOMOBILE EMISSIONS.)

Aldrin.   An agricultural insecticide which, together with other CHLORINATED HYDROCARBONS such as Endrin, DDT, Dieldrin, and benzene hexachloride, are major and serious pollutants. They have all been found in significant quantities in the milk of human mothers in the United States.

In October 1974 both Aldrin and Dieldrin were banned by the US government because of strong evidence that they are powerful CARCINOGENS. The ban will continue until a final decision is reached on whether or not to prohibit production and use of the chemicals permanently.

United States Food and Drug Administration statistics show that 83 per cent of all American-produced dairy products, 88 per cent of garden fruits and 96 per cent of meat, fish and poultry contain traces of Dieldrin. Furthermore, 99·5 per cent of human tissues taken during post-mortems in a 1971 study contained an average of 0·29 parts per million (ppm) of Dieldrin. The US Agriculture Department has warned that it will ban the sale for human consumption of meat containing more than 0·3 ppm. The use of Aldrin and Dieldrin has been restricted in Britain since 1960.

Algae. Extremely simple unicellular or multicellular plants which utilize the process of PHOTOSYNTHESIS for life. Most of them thrive in a wet environment (fresh water or marine) such as lakes, rivers and damp walls. They can cause problems in lakes and reservoirs if there is an excess of NUTRIENTS, as their excessive multiplication results in an algal 'bloom' and when they die the decay process is most unpleasant. Lakes that are rich in nutrients and as a consequence are highly productive in algae and other organic matter are said to be eutrophic.

A novel proposal is that algal blooms resulting from EUTROPHICATION should be harvested and fed to organisms higher up the FOOD CHAIN that have economic value, such as oysters. Others have suggested that algae should be cultivated in suitable nutrient liquids to produce vegetable protein.

Algae also form one of the constituent symbiotic partners in LICHENS. (♢SYMBIOSIS.)

Algal bloom. ♢ALGAE.

Alkali Inspectorate. The control of many of the major industrial processes in the United Kingdom capable of polluting emissions is vested in HM Alkali and Clean Air Inspectorate, which enforces the Alkali Etc. Works Regulation Acts in England and Wales, and in HM Industrial Pollution Inspectorate for Scotland. Currently the Alkali Inspectorate is administered as part of the Health and Safety Executive. (♢MEANS, BEST PRACTICABLE.)

Alpha particle (α particle). An alpha particle has a positive charge, consists of two protons and two neutrons (in effect the nucleus of a helium atom) and is emitted from the nucleus of an atom. A nucleus can spontaneously emit an alpha particle when its mass is greater than the combined masses of the product or daughter nucleus and the alpha particle. Spontaneous alpha emission takes place with many of the nuclei heavier than lead.

Alpha particles cause high ionization and are large in comparison with other radiating particles. They therefore lose energy very quickly. They have little penetrating force, 0·001–0·007 centimetres in soft tissue, but once inside the body, either by inhalation or through a wound, they are biologically more damaging.

Alpha radiation. A stream of fast-moving alpha particles emitted from the nuclei of radioactive elements. They are easily absorbed by matter. (♢RADIONUCLIDE.)

Amaranth. The most common red food colour in both the UK and the USA until it was banned there in 1976 after research had shown that large doses had caused cancer in rats. Amaranth is still used in the UK but is gradually being replaced by Allura Red.

Amino acid. Essential component of PROTEINS, consisting of amino (NH_2) and acidic carboxyl (COOH) groups. There are some 20 different amino acids normally present in proteins and a balanced diet must consist of the right intakes of these acids. There are eight essential amino acids that can only be obtained from the environment by heterotrophic means. The other non-essential amino acids can be manufactured by the human organism usually from the essential ones. 'First class' protein is the name given to protein that contains these eight essential amino acids and man's daily need is estimated at 30–70 grammes, some of which should preferably come from animal sources.

Anaerobic process. Any process (usually chemical or biological) that is carried out without the presence of air or oxygen, e.g. in a watercourse that is heavily polluted with no dissolved oxygen present. The anaerobic decay processes produce methane (CH_4) and hydrogen sulphide (H_2S) which is an evil-smelling gas.

Some anaerobic organisms, e.g. the denitrifying bacteria, are poisoned by the presence of oxygen. ($\Diamond$ AEROBIC PROCESS.)

Anaerobic treatment. Although AEROBIC PROCESSES are almost always used for the reduction of BIOCHEMICAL OXYGEN DEMAND (BOD) in organic effluents, interest in ANAEROBIC PROCESSES is growing, particularly for high BOD wastes from industry.

One such application has been reported (*Process Engineering*, May 1975, p. 6) on waste containing a BOD of 12 000 milligrammes per litre resulting from a starch and sugar content. The process is shown in Figure 2 and uses a digester which must be kept heated to allow sufficient bacterial activity. The result is methane, which is a source of light and heat. Between 85 and 95 per cent BOD is removed in this way, which means that the effluent from the digester would normally require aerobic treatment before final discharge.

The advantages that are claimed for the anaerobic process are that the plant is much smaller than that for an aerobic process, and the volume of sludge produced can be one-tenth of that from aerobic plants; it is said to be less objectionable to handle. It is extremely suitable for high BOD effluents. ($\Diamond$ EFFLUENTS, PHYSICO-CHEMICAL TREATMENT OF.)

Anchovy fisheries. The anchovy is a small fish of the herring family which can be easily converted to fish-meal. The major source of anchovy

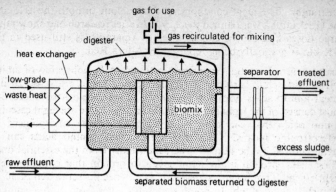

Figure 2. Anaerobic effluent treatment.

is the Peruvian fishing industry, which became the world's largest producer in the early 1960s. Approximately 96 per cent of the anchoveta catch from the Peruvian (Humbolt) current was converted into fish-meal. It represents the world's largest ocean harvest of a single species, and its export effectively deprives the South American continent of 50 per cent (1965–7) more protein than it is producing as meat, including Argentinian deliveries to Europe and North America. (⟡ FOOD FROM THE SEA; MAXIMUM SUSTAINABLE YIELD; WHALE HARVESTS.)

Angiosarcoma. A rare form of cancer commonly associated with the liver. (⟡ POLYVINYL CHLORIDE.)

Antagonism. A state in which the presence of two or more substances diminishes or decreases the toxic effects of the substances acting independently. It is the opposite of SYNERGISM.

Antibiotic. A chemical substance, produced by micro-organisms' which has the capacity, in dilute solution, to inhibit the growth of or destroy bacteria. Originally derived from moulds or bacteria, antibiotics are used to treat and control infectious diseases in man, animals and food crops, to stimulate the growth of animals, and to preserve foods.

With the large-scale use of antibiotics, particularly in farming (⟡ MODERN FARMING METHODS), the emergence of resistant bacteria has become an increasing problem. As resistance develops, an antibiotic becomes less effective and dosages must be increased. Particularly disturbing is the mechanism of transferable drug resistance, by means of which one species of bacteria previously susceptible to a certain antibiotic suddenly becomes resistant to it by virtue of cell-to-cell contact with a species of bacteria which is already resistant to the antibiotic. This resistance can be acquired to several drugs simultaneously.

For a full discussion of the dangers associated with the indiscriminate use of antibiotics in agriculture, see M. Taghi Farvar and J. P. Milton (eds.), *The Careless Technology*, Tom Stacey Ltd, 1973.

Aquifer.   An underground water-bearing layer of porous rock, e.g. sandstone, through which water can flow after it has infiltrated the upper layers of soil ($\Diamond$INFILTRATION). The London Chalk Basin is a typical example of such an aquifer; it supplies three million cubic metres of water per day. Water may also be extracted from the chalk under the Cotswolds to meet the expected demand from London by the end of the century.

Aquifer management and RIVER REGULATION are now practised conjunctively in the UK to maximize the amount of water available for consumption from rainfall over a particular catchment area. ($\Diamond$WATER SUPPLY.)

Arable land, Potential of.   In 1967 the President of the United States Scientific Advisory Committee reported the potentially arable land on earth to be 7·8 thousand million acres, which is three *times* the area actually harvested. The majority of this land is in Africa, Asia and South America.

At first sight, therefore, prospects look bright indeed. A closer look suggests that caution is advisable. About 4·13 thousand million acres (more than half the estimated total) lie in tropical areas where, in spite of abundant rainfall, soils are very poor and unstable once the forest is cleared. Most of the potentially arable land in Asia (1·55 thousand million acres) would not support large-scale agricultural development without irrigation – and the costs of producing such irrigation would be huge ($\Diamond$DESALINATION).

In fact, most of the land that can be *economically* harvested is already under cultivation. Given the will to invest capital on a large scale, what sums are we talking about? Ehrlich quotes the following example:

Under the optimistic assumption that one acre of land will support one person, and the even more optimistic assumption that development costs will be only $400 per acre (the cost of irrigating alone now averages almost $400 per acre), the world would have to invest $28 *billion per year* simply to open new lands to feed the people now being added to the population annually (P. R. Ehrlich and A. H. Ehrlich, *Population, Resources, Environment*, W. H. Freeman, 1970, p. 0).

Arithmetic–geometric ratio (A/G ratio). A principle used by geologists for estimating the abundance of certain types of ore deposits, the idea being that as the grade of the ore decreases arithmetically its abundance increases geometrically until the average abundance in the earth's crust is reached. This has led certain economists to assert that the problem of non-renewable mineral resources is a non-problem, since as demand increases mining will simply move to poorer and poorer ores, which are assumed to be progressively more and more abundant.

As one would expect, the facts are against such a simplistically reassuring view of the situation. The A/G ratio is applicable only to a very limited number of ores, and only within certain limits. In addition, although the economic costs of working lower and lower grade ores might be absorbed in some cases, the energy costs could not (⊳ENERGY DEMAND).

Asbestos. A collective term for a group of magnesium silicate materials ($MgSiO_4$). It is a fibrous mineral of variable length with excellent resistance to fire, heat and chemical attack. It is widely used in building products, gaskets, brake linings, and roofing materials.

There are two main forms of asbestos:

'Blue asbestos' or crocidolite, which is so dangerous that its use is virtually banned now in the UK. A survey of male asbestos workers by the TUC Centenary Institute of Occupational Health suggests that 30 years after first exposure about one in 200 workers will be found to have died from mesothelioma (malignant tumours) associated with blue asbestos. Blue asbestos is still present in old buildings, boiler plant, ships, etc., and its presence is a substantial threat to any demolition worker.

'White asbestos' or chrysotile, which can be readily spun or woven into tape. (Other varieties are amosite and tremolite.)

The dangers associated with asbestos are grave. Asbestosis, a scarring of the lung, cancers of the bronchii, pleura and peritoneum may result from breathing the minute fibres. Asbestosis may result from exposures as short as six weeks in heavy dust concentrations. Brief exposure to blue asbestos can manifest itself later in life with fatal results.

In the UK, the 1931 Asbestos Regulations put the onus squarely on the employers to ensure that asbestos dust was removed *at source*, respirators being regarded as only secondary protection. Astonishingly, these regulations were apparently disregarded by employers, with a resulting high record of sickness and death in the industry. The Factory Inspectorate, which had the responsibility for ensuring that the factory acts relating to asbestos were properly enforced, appear to have been completely ineffective, with a total record of three prosecutions in 30 years.

The current legislation is under Asbestos Regulations 1969. Its recommendations are:

1. Substitute material to be used where possible.

2. Exhaust ventilation to reduce atmospheric contamination to below the THRESHOLD LIMITING VALUE (TLV). Monitoring is mandatory and enforced by the Factory Inspectorate.

The current TLVs for asbestos dust are: chrysotile, amosite and tremolite – 2·0 fibres per cubic centimetre of air; crocidolite – 0·2 fibres per cubic centimetre of air, where the fibres measured are five millionths of a metre in length or greater. The sample is to be taken by a prescribed membrane filter which is then scanned by a microscope at 400–450 magnification.

It would be optimistic in the extreme to assume that these stringent precautions are obeyed everywhere, particularly in the demolition and insulation industries.

This material is a major hazard to health and no concentration of asbestos dust may be presumed safe.

Aspergillus niger (AN). A fungus capable of breaking down vegetable matter (e.g. carob pods, which come from the locust tree which is extensively grown in many developing countries) to produce SINGLE CELL PROTEIN (SCP). The strain used is called M1 and is capable of doubling its weight in five to ten hours, provided inorganic nutrients are added.

A village-level technology based on this process is being developed to use carobs and other starchy vegetable matter as a source of protein for pigs and cattle.

Asphyxiating pollutants. Asphyxiation is deprivation of oxygen, either through obstruction of the air passages or, as in the case of CARBON MONOXIDE, the inability of the blood to carry oxygen. HYDROGEN SULPHIDE (H_2S) is an irritant at very low concentrations, but it can also paralyse the respiratory system at slightly higher concentrations causing death by asphyxiation.

See G. L. Waldbott, *Health Effects of Environmental Pollutants*, C. V. Mosby, St Louis, 1973.

Aswan Dam. ⇨DAM PROJECTS.

Atmosphere. The gaseous envelope around the earth which contains by volume nitrogen 78 per cent, oxygen 21 per cent, carbon dioxide

0·03 per cent, plus inert gases such as argon 0·93 per cent and helium 0·0005 per cent. The composition of our present atmosphere is the result of the CARBON CYCLE and the NITROGEN CYCLE and the atmosphere is renewed and maintained by these processes.

The atmosphere, like any other natural resource, is finite and 99 per cent of its mass is within 20 miles of the earth's surface. It is in contact at its inner edge with land and water and changes in the atmosphere can induce changes in land or water such as the amount of rainfall on a particular area. The changes can be direct or they can be complex chain reactions.

The role of the atmosphere in the earth's radiation balance is unique. The incoming solar radiation is absorbed by the earth and reradiated into outer space as long-wave radiation, but the two processes are not in balance except over, say, a year when the total incoming radiation may balance the total outgoing. A very small change in either the out-going or incoming radiation can have very large effects. Hence, the concern for the carbon dioxide released by combustion of fossil fuels (◊GREENHOUSE EFFECT), the DUST particles released by combustion processes, the effects of AEROSOL PROPELLANTS on the ozone layer, or supersonic aircraft (◊CONCORDE). (◊◊THERMAL POLLUTION; ALBEDO; CLIMATE.)

Atom. The smallest part of an ELEMENT that can take part in a chemical reaction. It consists of a core, called the NUCLEUS, surrounded by negatively charged ELECTRONS which move in orbits around it.

Atomic number. The number of ELECTRONS rotating around the NUCLEUS of an ATOM. This determines the chemical behaviour of an atom. The atomic number is constant for each ELEMENT and its ISOTOPES.

Automobile emissions. Generic name for the emissions from car exhausts, the principal components of which are lead, nitrogen oxides (NO_x), unburnt hydrocarbons, water vapour, carbon monoxide (CO), aldehydes, and carbon dioxide (CO_2).

Carbon monoxide, lead and nitrogen oxides are major pollutants in our cities, due to increased automobile emissions resulting from an increase in the number of cars on the road and the use of high-powered engines. They have significant health effects, particularly under SMOG and PHOTOCHEMICAL SMOG conditions. In busy streets the carbon monoxide concentration can rise to 15–20 parts per million (ppm); over 100 ppm have been measured in London. The background concentration of carbon monoxide in the atmosphere is less than 0·1 ppm.

The high compression ratios required for high-performance engines need fuels with 'anti-knock' properties. This is achieved by adding tetraethyl and tetramethyl to the fuel, and a large part of the lead – which is toxic – is emitted from the car exhaust in a fine particulate form that can be readily inhaled. Lead also poisons the catalytic reactors that are beginning to be used in car exhaust systems to minimize emissions in the USA.

Photochemical smog is triggered by the emission of nitrogen oxides, and the amounts emitted from car exhausts have increased as compression ratios, and hence engine temperatures, have increased, particularly in urban areas. It is estimated that in the United States from 1946 to 1968 total emissions of nitrogen oxides increased sevenfold.

In California legislation has been enacted to effect reductions of 87 per cent for carbon monoxide, 95 per cent for hydrocarbons, and 75 per cent for oxides of nitrogen. This can be achieved by means of a catalytic reactor in the car exhaust system which converts the unburnt hydrocarbons to carbon dioxide and water, and the carbon monoxide to carbon dioxide. The nitrogen oxides can also be reduced. Recirculation of exhaust and crankcase vapours back into the combustion chamber can also be carried out to further reduce emissions. The use of lead in petrol is also to be severely curtailed.

Automobiles. The automobile is of great benefit in the mobility and freedom that it has given to millions of people. We are reaching the stage, however, where in the city the disadvantages outweigh the advantages. That such a situation should have come about is in part the fault of the major car producers, who have increased engine power, in the United States, to ludicrous extremes.

In the first place, the energy demands of 200-odd million cars, each burning around its own mass in fuel every year with between 10 and 20 per cent efficiency, are quite enormous. Cars consume about 6 per cent of the total world energy supply, and this monstrous fleet of vehicles is officially predicted to double by 1985 (Leach, 1972). This prediction may not be realized due to the world-wide rise in oil prices.

Two-thirds of this consumption occurs in the United States where automobiles are twice as big as anywhere else and where, ironically, legislation is tending to increase fuel consumption: legislation aimed at improving safety is tending to produce heavier cars which require higher performance engines; legislation aimed at reducing exhaust emissions reduces the power available from a given engine size, thus tempting manufacturers to increase engine size even further.

The total energy cost of automobiles is, of course, much larger and includes the energy used in the construction of roads, factories, retail and service stations, etc. Input–output analysis of US energy consumption in 1963 put the grand total for cars at over 20 per cent (Herenden, 19).

Secondly, the internal-combustion engine has become a very efficient smog-generator. High-powered engines running at low speeds, as they are forced to do in congested streets, are a major source of urban pollution from carbon monoxide, lead and smog ($\diamond$AUTOMOBILE EMISSIONS).

The adverse effects of the automobile are not restricted to pollution. Their effects on our cities are incalculable. They are undoubtedly contributing to city-centre 'rot', with the more affluent moving to the suburbs – where the car becomes essential in the absence of a properly planned mass-transit system.

Furthermore, the development of road systems for the private car has made road haulage more profitable, with the result that trucks carry more of the freight previously hauled by the railways. For the same freight haulage, trucks burn nearly six times the fuel as railways, and though financial savings may be claimed for trucks, energy savings cannot.

Properly designed vehicles, capable of carrying four people in comfort at adequate speeds (say 40 to 45 mph) with fuel consumption between 100 and 200 miles per gallon are well within the capability of current technology. Such vehicles could be adapted to work on alternative fuels, such as methanol (methyl alcohol), methane, hydrogen and ethanol. No doubt such vehicles will eventually be produced. Whether the inevitable transition to such ecologically acceptable vehicles can be carried out without traumatic upheavals and massive unemployment in the car industry depends upon the vision and foresight of the captains of this particular industry. Judging by past and present performance, the omens are not propitious.

G. Leach, *The Motor Car and Natural Resources*, OECD, Paris, 1972.
R. A. Herenden, Center for Advanced Computation Document No. 69, University of Illinois, Urbana, 19 .
See also A. Aird, *The Automotive Nightmare*, Arrow Books, 1974.

Autonomous house; also known as the Ecohouse. The self-contained dwelling where man simulates the ways of ECOSYSTEMS, i.e. his wastes are converted to fuel by anaerobic digestion for METHANE production and the residues from the digestion are used for growing food ($\diamond$ANAEROBIC TREATMENT). The food residues are composted and/or used for methane production. Solar energy is trapped by the GREENHOUSE EFFECT and used for house, crop and water heating. A windmill would be used for electricity. Thus, given sufficient space, sunshine, rainfall and wind, the autonomous house is in theory a self-contained system recycling its own wastes and using the sun as its energy input.

Much research and development needs to be done before this can be achieved. The practitioners still require an initial energy input to build the house, provide the raw materials such as glass, plastics, paint, etc., but the concept may lead to much more energy-efficient housing.

Autotrophic organism. An organism that does not require organic material for food from the environment but can manufacture food from inorganic chemicals. For example, chlorophyll-containing plants or algae can manufacture their food from water, carbon dioxide and nitrates using the sunlight for an energy source (◇PHOTOSYNTHESIS). HETEROTROPHIC ORGANISMS depend on the activities of the autotrophic ones for food.

Azodrin. ◇ORGANOPHOSPHATES.

B

Background concentration of pollutants. If the atmosphere in a particular area is polluted by some substance from a particular local source, then the background level of pollution is that concentration which would exist without the local source being present. Measurement would then be required to detect how much pollution the local source is responsible for.

Sometimes the word 'background' is used to mean the concentration of the substance some distance from the particular source and therefore largely uninfluenced by it.

The term is also used in radiation work to mean the background level of radiation from natural sources or from sources other than that being measured (⟡IONIZING RADIATION, EFFECTS).

Bacteria. Class of organisms which do not possess CHLOROPHYLL. They usually multiply rapidly, by division.

Bacteria occur virtually everywhere in the BIOSPHERE and are responsible, for example, for the souring of milk or the decay of dead animals. They participate in the CARBON CYCLE, can fix nitrogen in the NITROGEN CYCLE, and decompose sewage (⟡SEWAGE EFFLUENT TREATMENT). As DECOMPOSERS they are responsible for the decomposition of organic matter into simple substances which can then be reincorporated into the biological cycles. They can also cause and transmit diseases, e.g. TYPHOID, tuberculosis, etc.

Bacteria, Use in mining. ⟡SOLUTION MINING.

Ballast voyage. The journey an oil tanker or other vessel makes with sea water carried in the oil-tanks as ballast to provide essential stability on the empty leg of its journey. The ballast is usually discharged at sea along with any residual oil from the tanks and consequently the practice is a major contributor to oil pollution (⟡OIL SLICKS). It is estimated that 20 million tonnes of oil were dumped on the world's seas in 1973 as a result of ballast-blowing. World oil consumption is about 3000 million tonnes per year, so less than 1 per cent is dumped at sea, but this is still an awful lot of oil which could cause marine pollution on a large scale.

Benzene hexachloride. ▷CHLORINATED HYDROCARBONS.

Berylliosis. ▷BERYLLIUM.

Beryllium (Be). Non-corrosive silver-white metal with very good resistance to both heat and stress. Its lightness and hardness find a wide variety of applications, especially in rocketry, aircraft, nuclear reactors. The main environmental sources are beryllium refining, alloying and fabricating operations and in the burning of coal.

Acute poisoning can result from exposure to airborne concentrations as low as 20 microgrammes per cubic metre for less than 50 days. The lung tissue is inflamed and fevers, chills and shortness of breath can become chronic from such short exposures.

In the UK the maximum allowable concentration at discharge has now been set at 0·1 microgrammes per cubic metre. It has been classified as a HAZARDOUS POLLUTANT in the USA.

Best practicable means. An essentially pragmatic philosophy for the control of emissions from scheduled processes. The term was first used in the 1906 Alkali Act and is now frequently referred to by the UK ALKALI INSPECTORATE who administer the legislation and who are responsible for controlling emissions from scheduled industrial premises (oil refineries, chemical works, cement plants, etc.).

The ideology behind best practicable means causes much controversy as it is considered to be a loophole which may allow industry to emit noxious or injurious substances in greater amounts than would have been allowed had absolute standards been imposed. Basically, it implies that, while better emission standards may be obtainable, industry must not be unduly penalized in its operations as the costs of pollution are offset by the social and economic benefits of a thriving industrial sector. Thus, the effective emission standards may have surprisingly wide tolerances depending on local environmental factors and in practice may often lag behind those that technologically are quite feasible.

Beta particles (β particles). Beta particles are electrons and therefore negatively charged. They are sparsely ionizing and have little penetrating ability, although more so than an alpha particle. Many atoms heavier than lead decay spontaneously by the emission of beta particles; however, a few emit positive particles (positrons) and these are also included under the term beta particle. Positrons are equal in mass and charge to an electron.

In negative beta-particle emission, a neutron in the original nucleus breaks up into a proton and an electron; the latter is emitted. This decay will take place in nuclei which have too many neutrons for complete stability. Conversely, in positive beta-particle emission there are too many original protons and these break up into neutrons and positrons.

Also emitted in beta radiation are neutrinos, which have a zero charge and rest mass.

Beta radiation. A stream of beta particles, i.e. electrons or positrons emitted from radioactive nucleus, possessing greater penetrating power than ALPHA RADIATION. (⟡RADIONUCLIDE.)

Bilharziasis. ⟡SCHISTOSOMIASIS.

Biochemical oxygen demand (BOD). A standard water-treatment test for the presence of organic pollutants. This biochemical test depends on the activities of bacteria and other microscopic organisms which in the presence of oxygen feed upon organic matter. The result of the test indicates the amount of dissolved oxygen in grammes per cubic metre used up by the sample when incubated in darkness at 20°C for five days. This in turn gives a measure of the quantity of organic material present.

The factors which influence the test are:

1. The presence of toxic substances which may inhibit the growth of micro-organisms during the test.

2. The chemical uptake of oxygen by substances such as ammonia or nitrites.

3. A departure from the specified incubation temperature of 20°C.

4. The availability of dissolved oxygen and nutrients.

The BOD test may be supplemented by chemical oxygen demand (COD) tests, but these do not measure the biodegradable matter which is of central interest (⟡DISSOLVED OXYGEN). Both tests are commonly run when a 'new' effluent is encountered.

Typical BOD and COD values exerted by organic wastes are shown in the table on p. 38. The COD column is the chemical oxygen demand, which uses potassium dichromate and oxidizes the organic matter completely and assesses the oxygen demand from materials that cannot be treated biologically.

Both the BOD and COD tests are important for effluent monitoring. The BOD test requires five days whereas the COD test can take only

Type of waste	BOD 5 days g m^{-3}	COD
Abbatoir	2600	4150
Brewery	550	—
Distillery	7000	10000
Domestic sewage	350	300
Pulpmill	25000	76000
Petroleum refinery	850	1500
Tannery	2300	5100

two hours. Thus, once an effluent BOD–COD ratio has been determined, the monitoring of the effluent by the COD test automatically establishes the BOD, as BOD–COD ratios are relatively constant for particular effluents, but of course vary from one waste to another.

Biodegradation. The ability of organic materials, i.e. those which come from living sources – e.g. plants, animals, to be decomposed by biological means. For example, the action of BACTERIA reduces the carbon of plant systems to carbon dioxide and water ($\lozenge$ CARBON CYCLE).

Non-biodegradable organic substances, i.e. chemical compounds containing carbon, hydrogen, oxygen and perhaps other elements, emanate from the synthetic chemicals industry and are not necessarily harmful to the environment. Plastics, for example, are visually undesirable as litter, but they are not environmentally hazardous whereas the organochlorine class of chemicals is potentially dangerous. ($\lozenge$ PLASTICS, DEGRADATION; ORGANOCHLORINES.)

Biological concentration. The mechanism whereby filter feeders such as limpets, oysters and other shellfish concentrate HEAVY METALS or other stable compounds present in dilute concentrations in sea or fresh water. One extreme example is that of limpets collected from rocks in the Severn Estuary which contained 550 parts per million of CADMIUM of which the ingestion of 50 milligrammes could be lethal. The limpet also contained 900 ppm zinc. The Severn Estuary is near a large zinc smelter and it is not improbable that the high concentrations resulted from industrial activity.

It should be noted that fungi, algae and bacteria which thrive on effluents are being considered for new sources of protein and these organisms can all concentrate biologically by factors of up to 10000. ($\lozenge$ PCBs.)

B. Silcock, 'Limpets reveal metal poison danger', *Sunday Times*, 6 June 1971.

Biological systems. ⬦FOOD CHAIN; CARBON CYCLE; NITROGEN CYCLE.

Biomass. The mass of living organisms forming a prescribed population in a given area of earth's surface. It is usually expressed in grammes per square metre ($g\ m^{-2}$).

Biome. A large, well-defined biological community, e.g. tundra, grassland, coral atoll. The biome has a particular form of vegetation and associated animals which have become adapted to the local conditions. In other words, it is a balanced ecological community.

Biometallurgy. The application of microbiological processes to mineral treatment especially for the extraction of non-ferrous metals from sulphide ores. (⬦SOLUTION MINING.)

Biophotolysis. The storage and use of electrons produced in the first stages of PHOTOSYNTHESIS which can then be used to make free hydrogen. This is another research route which investigates the use of energy from the sun to produce electricity or fuels such as hydrogen, in this case directly. Experimental work on algal systems has shown that this can be done in the laboratory.

Biosphere. The region of the earth and its atmosphere in which life exists. It is an envelope extending from up to 6000 metres above to 10000 metres below sea level and embraces alpine plant life and the ocean deeps. The special conditions which exist in the biosphere to support life are: a supply of water; a supply of usable energy; the existence of interfaces, i.e. areas where the liquid, solid and gaseous states meet; the presence of nitrogen, phosphorus, potassium and other essential nutrients and trace elements; a suitable temperature range; and a supply of air (although there are anaerobic forms of life).

Biospheric cycles. To maintain the biosphere and thus life, essential materials must be recycled so that after use they are returned in a reuseable form. In order to sustain this recycling a supply of energy is required – solar energy.

The principal elements in the biosphere are nitrogen, oxygen, hydrogen and carbon. As these elements constitute 99 per cent of all living things, including plants, they are obviously of prime importance. Other elements are also important, e.g. phosphorus, potassium, sulphur and calcium, and all have their own renewal cycles, so that they become available for reuse.

The major biospheric cycles are the CARBON CYCLE, the OXYGEN CYCLE, the NITROGEN CYCLE, and the water or HYDROLOGICAL

CYCLE. They are all interlinked and operate simultaneously and maintain all life on earth.

Biotic index. The diversity of species in an ECOSYSTEM is often a good indicator of the presence of pollution. The greater the diversity, the lower the degree of pollution. The biotic index is a systematic survey of aquatic organisms which is used to correlate with river quality. It is based on two principles:

1. Pollution tends to restrict the variety of organisms present at a point, although *large numbers* of pollution-tolerant species may persist.

2. In a polluted stream, as the degree of pollution increases, key organisms tend to disappear in the following order:
Plecoptera (stone fly), *Ephemeroptera* (mayflies), *Trichoptera* (caddis fly), *Gammarus* (freshwater shrimp), *Asellus*, bloodworms, tubificid worms.

The biotic index ranges from 0 to 10. The most polluted aquatic environment which therefore contains the smallest variety of organisms is at the lowest end of the scale, the clean streams are at the highest.
Streams can in practice have a range of biotic indices. As a rule of thumb a stream with a biotic index above 6 would support fish; below 4 it will not and 1 or less is very toxic with probably only tubificid worms present.

Birth control. The conscious control of POPULATION GROWTH by the deliberate use of CONTRACEPTION and other 'unnatural' methods (◇ ABORTION). It took all of history for the world's human population to reach 1000 million in 1850, only 80 more years to reach 2000 million (in 1930). At the present growth rate (70 million a year) we shall have doubled again – to 4000 million – in seven or eight years, the doubling time having reduced to 52 or 53 years. At this rate, the next doubling time – to 8000 million – will occur in only 35 years.
Can birth control be effective in ensuring a levelling off of population growth to a sustainable steady state? Although essential, it is slow to take effect. If birth rates were, by some miracle, cut instantaneously to two children per couple, the world's population would still rise inexorably for another forty years or so for the simple reason that such a large proportion of our present population is young, particularly in the underdeveloped countries. The next generation's mothers are already alive (◇ AGE-SPECIFIC BIRTH AND DEATH RATES). And in any case, birth-control programmes are having only a limited success. Many governments in the third world are profoundly suspicious of such programmes and some, like Argentina's, are actively hostile, as is the

Vatican. At the 1974 United Nations Conference at Bucharest both helped stage-manage an extraordinary coalition that wrecked an already feeble world plan. Many third-world countries are convinced that birth-control programmes are a subtle trick to prevent them becoming more affluent.

In spite of these facts, a large body of religious opinion is opposed to any form of 'artificial' birth control on moral grounds. The present Catholic attitude to birth control is the subject of great debate within the Church. A large and growing proportion of Catholics are critical of Pope Paul's 1968 encyclical *Humanae Vitae*, which condemned the use of contraceptives. It is already widely ignored by large sections of the Catholic community – in general the more well educated and affluent.

Even where contraception is widely available, as in India, results have been disappointing. In UNDERDEVELOPED COUNTRIES large families are not so much due to religious beliefs as economic pressure, male children being regarded as a form of economic security. China alone seems to be having some success on a national scale with a planned programme of birth control. Nevertheless, on the World Bank's conservative estimate, the earth will have another 1200 million people by 1990.

Birth rates. The birth rate is usually expressed as the number of babies born per thousand people per year. The total number of births during the year is divided by the estimated population at the *midpoint* of the period. Estimates of death rates are calculated in the same way, i.e. total number of deaths during the year divided by the estimated population at mid-year and expressed as deaths per thousand people. (◊POPULATION GROWTH; AGE-SPECIFIC BIRTH AND DEATH RATES.)

Blue-green algae. Algae which, in addition to chlorophyll, possess red and blue pigments. Some are capable of fixing nitrogen directly and so make it available for plants and animals; others convert nitrates to nitrites. They are, therefore, involved in the NITROGEN CYCLE.

Boreholes. ◊WELLS.

Bottle. Glass or plastic container, which can be returnable or non-returnable. The ratio of non-returnable to returnable bottles in the UK is 5:1.

Bottles come in different colours, shapes and sizes. Since conservationists argue strongly against the throw-away philosophy, with some justification on ENERGY consumed, a strong case can be made for bottle standardization and the enforced use of returnables so that both energy and materials are conserved. (◊RECYCLING; TRIPPAGE.)

Breeder reactor. ⇨NUCLEAR REACTOR DESIGNS; NUCLEAR RE-
ACTORS, CLASSES.

Burn out (1). A measure of the performance of a domestic refuse
incinerator; high burn out means an ash with low content of fixed
carbon and an absence of putrescible material.

Burn out (2). The failure of a nuclear reactor heat-exchange surface
which can result in the loss of heat-exchange fluids and may in very
extreme circumstances result in the release of radioactive products. It
is equivalent to 'melt out' in American terminology.

C

Cadmium (Cd). A soft silvery HEAVY METAL, atomic weight 112·4. It is used in semiconductors, control rods for nuclear reactors, electroplating bases, PVC manufacture, and batteries. World production is around 18000 tonnes per year produced in conjunction with zinc smelting.

Cadmium serves no biological functions, is toxic to almost all systems, and is absorbed into the human organism without regard to the amount stored. Very small doses can cause severe vomiting, diarrhoea and colitis; pneumonia can develop as a consequence. Poisoning has developed from sources such as home-made punch stored in a cadmium-plated bowl. Changes in the heartbeat rate of persons in atmospheres containing as little as 0·002 microgrammes per cubic metre have been reported (maximum allowable concentration is 100 microgrammes per cubic metre). Hypertension is a common complaint of those who have ingested the metal at doses well below those regarded as toxic. Long exposure in lead and zinc smelting areas in Japan led to the so-called Ouch-ouch (Itai-itai) disease where the sufferers had severe joint pains and eventual immobility as a result of skeletal collapse. Cadmium leads to bone porosity and inhibition of bone-repair mechanisms.

Caesium–137. Radioactive isotope of caesium, produced in nuclear reactors as a fission product of uranium. It has a half-life of 33 years. It is chemically similar to potassium. When absorbed into the body, it is not concentrated like STRONTIUM–90 but spread throughout the muscles.

Calcium deficiency. DDT and other chlorinated hydrocarbons such as polychlorinated biphenyls severely interfere with a bird's ability to metabolize calcium. The result is the production of egg shells which are so thin that they are crushed by the weight of the nesting birds.

Calorie (cal). Physicists define the calorie as the heat required to raise 1 gramme of water 1°C. This is a minute quantity of heat in practical

43

terms. The term used on diet sheets refers to the KILOCALORIE, which is 1000 times larger.

Cancer. A disordered growth of cells which can invade and destroy body organs and thereby cause incapacity or death. It can be initiated by CARCINOGENS. The many causes of cancer are subjects of major medical research. (⬦ASBESTOS; HEAVY METALS; IONIZING RADIATION.)

Cancer due to air pollution. ⬦CARCINOGEN.

Cancer due to amaranth. ⬦AMARANTH.

Cancer due to asbestos. ⬦ASBESTOS.

Cancer due to chloroprene. ⬦CHLOROPRENE.

Cancer due to chrome waste. ⬦CHROME WASTE.

Cancer due to DES. ⬦DIETHYLSTILBOESTROL; FEED–CONVERSION RATIO.

Cancer due to hair dyes. ⬦MUTAGEN.

Cancer due to nitrate preservatives. ⬦NITRATE PRESERVATIVES.

Cancer due to organophosphates. ⬦ORGANOPHOSPHATES.

Cancer due to pesticides. ⬦ALDRIN.

Cancer due to PVC. ⬦POLYVINYL CHLORIDE.

Carbohydrates. ⬦NUTRITIONAL REQUIREMENTS, HUMAN.

Carbon cycle. The atmosphere is a reservoir of gaseous carbon dioxide but to be of use to life this must be converted into suitable organic compounds, i.e. fixed, as in the production of plant stems. This is done by the process of PHOTOSYNTHESIS. The productivity of an area of vegetation is measured by the rate of carbon fixation.

For example, tropical rain forest growth will fix 1 to 2 kilogrammes per square metre per year, whereas barren areas such as tundra will fix less than 1 per cent of this amount. The carbon fixed by photosynthesis is eventually returned to the atmosphere as plants die and the dead organic matter is consumed by the decomposer organisms. Thus the global carbon cycle on land (there is a complimentary oceanic one as well) is as pictured in Figure 3.

In Figure 3, the atmosphere contains 700 thousand million (700×10^9) tonnes of carbon as carbon dioxide. The fixed amount of carbon in the biomass (plants and animals except the decomposers) is 450×10^9 tonnes and the dead organic matter reservoir is 700×10^9

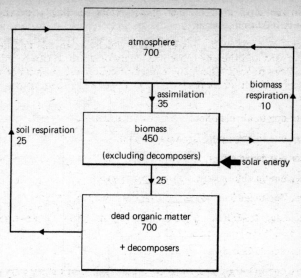

Figure 3. Carbon cycle on land. The quantities in Figures 3–5 are in thousand million (10^9) tonnes. The figures in the boxes show the amount of carbon either fixed or available for fixation. The figures with the arrows show the flow rates of carbon which is assimililated or respired per year.

tonnes. The land-based system is in a steady state, i.e. the inputs and outputs balance. The input has been estimated as 35×10^9 tonnes carbon per year from the atmosphere as carbon dioxide, of which 25×10^9 tonnes per year are fixed and 10×10^9 tonnes per year are returned to the atmosphere as carbon dioxide, this component being used as an energy source to keep the plants alive. The dead organic matter reservoir has an input of 25×10^9 tonnes carbon per year as fixed carbon which is decomposed to 25×10^9 tonnes per year of carbon as carbon dioxide. The system is in balance, powered by solar energy. In the process oxygen is also produced so that the carbon and oxygen cycles interlink.

Complementing the carbon cycle on land is the equivalent at sea. This differs from the land cycle in that carbon dioxide is soluble in sea water and is thus available for the phytoplankton. These are single-celled organisms that take up carbon dioxide and, by photo-synthesis, use it to produce carbohydrates and oxygen which also dissolves in the water, so the oceanic system is virtually closed. Further up the food chain, zooplankton and fish consume the carbon fixed by the phytoplankton and in turn die and decompose, as do the uneaten phytoplankton, so that the carbon dioxide is replaced. Less than 1×10^9 tonnes carbon per year are deposited to sediments. Thus the oceanic system is also in a steady state as shown in Figure 4. The

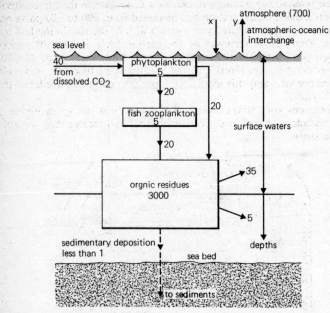

Figure 4. Oceanic carbon cycle.

inputs to the dead organic matter reservoir are 20×10^9 tonnes per year from fish and zooplankton and 20×10^9 tonnes per year from phytoplankton. This is balanced by an output of 35×10^9 tonnes per year as carbon dioxide to the surface layers and 5×10^9 tonnes per year carbon dioxide to the depths.

Now the mixing time for exchange between surface layers and the depths is roughly 1000 years and it is this mixing time which controls the balance between carbon dioxide in the atmosphere and the oceans. Thus any extra carbon dioxide from man's fossil fuel combustion activities will increase the atmospheric carbon dioxide concentration and, since the natural system is a closed cycle, the assimilation of excess carbon dioxide will take 1000 years before equilibrium is attained through atmospheric–oceanic interchange. As we cannot control the carbon (or any other) cycle, we should be careful not to disturb natural systems too far.

Man's fossil fuel combustion activities account for 5×10^9 tonnes of carbon per year input as carbon dioxide which is not wholly assimilated by the interchange mechanism and as a consequence the atmospheric carbon dioxide concentration has increased from 290 to 320 parts per million in the last hundred years with a fifth of the rise in the last ten years. The natural turnover of the carbon cycle is 75×10^9 tonnes carbon per year (40×10^9 tonnes per year from the ocean, 35×10^9 tonnes per year on land) and the man-induced turnover is 5×10^9 tonnes per year and this input of 6·5 per cent extra has displaced the equilibrium.

The composite carbon cycle in the biosphere is shown in Figure 5 and includes the additional fossil fuel input and atmospheric–oceanic interchange.

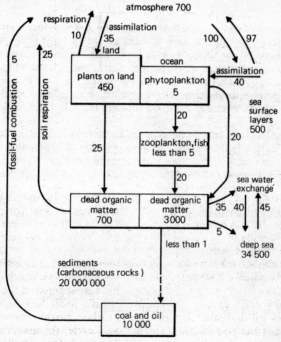

Figure 5. Decomposite carbon cycle.

Carbon dioxide (CO$_2$). Gas produced by the complete combustion of carbonaceous materials, by decay organisms such as aerobic DE-COMPOSERS, FERMENTATION, the action of acid on limestone. It is exhaled by plants and animals, and utilized in the CARBON CYCLE.

The combustion of fossil fuels has raised the atmospheric carbon dioxide concentration from 290 to 320 parts per million in the last century with most of the rise occurring in the last decade. The current net annual increase is estimated at 0·7 parts per million.

Carbon dioxide is responsible for the atmospheric GREENHOUSE EFFECT.

Carbon monoxide (CO). A colourless odourless gas, lighter than air, formed as a result of incomplete combustion. It is a chemical poison when inhaled, as it can penetrate tissues and is absorbed into the bloodstream where it combines with the haemoglobin of blood cells, thus displacing oxygen from the blood and depriving brain and heart tissues of oxygen. (⊹AUTOMOBILE EMISSIONS; SMOG; ASPHYXIATION.)

Carcinogen. Any compound or element which will induce CANCER in man or other animals. It is now accepted that a large proportion of human cancers are directly associated with environmental agents, in particular chemical and physical agents, e.g. some chlorinated hydro-carbon PESTICIDES on the one hand and ASBESTOS fibres on the other. Carcinogens can be inhaled, e.g. tobacco smoke and asbestos fibres; ingested; absorbed through the skin (insecticides, hair dyes).

A no-effect level cannot be assumed for carcinogens as the development of cancer is the result of an accumulation of irreversible cellular damage. In contrast, exposure to toxic substances can in many cases result in little or no damage below certain levels and they are eventually excreted from the human system.

Health Hazards of the Human Environment (World Health Organization, 1972) lists the following classes of environmental carcinogenic agents:

1. Polynuclear compounds. These occur in tobacco and coal smoke, exhaust gases; a major compound is benzo (α) pyrene (BP) found in all kinds of soot and smoke.

2. Aromatic amines. These are mainly found in industrial environments; they can also contaminate foodstuffs and plastics.

3. CHLORINATED HYDROCARBONS. A major group which embraces industrial solvents, DDT, Aldrin, Dieldrin, and other pesticides. Some

members of this family are notorious for their ubiquity, persistence and ability to accumulate in living systems.

4. N–nitroso compounds. These can be found in industrial solvents of chemicals. They are formed in the human intestine through bacterial action following the ingestion of nitrites.

5. Inorganic substances. The HEAVY METALS are included in this class, particularly beryllium, selenium, cadmium.

6. Naturally occurring agents. These are the so-called toxins and occur in spoiled food.

7. Hormonal carcinogens. Hormonal imbalances can induce cancer. The growth-promoting chemical DES (diethylstilboestrol) has been implicated as a carcinogen.

The known range is very extensive and any substance that acts as a MUTAGEN must be considered a potential carcinogen until proved otherwise.

Carnot efficiency. This is a direct outcome of the second law of thermodynamics and places the upper limit on the maximum possible conversion of heat to work in a heat engine. It is usually derived as:

$$\text{Carnot efficiency} = \frac{T_{source} - T_{sink}}{T_{source}}$$

where the TEMPERATURES are on the absolute scales. Thus, for a heat engine supplied with steam at 900 K and sink (e.g. a river at 300 K), the Carnot efficiency is

$$\frac{600}{900} = 0.66$$

or 66 per cent – the maximum possible conversion of heat to work. In practice, the actual efficiency is much lower. (◊LAWS OF THERMO-DYNAMICS; DIRECT ENERGY CONVERSION.)

Catalysis. The acceleration (or retardation) of a chemical reaction by a relatively small amount of a substance (the catalyst), which itself undergoes no permanent chemical change, and which may be recovered when the reaction has finished.

Catalyst. A substance or compound that speeds up the rate of chemical or biochemical reactions. Catalysts used by the chemical industry are usually metals, such as vanadium which speeds up synthesis reactions. ENZYMES are organic catalysts.

Catchment area.   The natural drainage area for PRECIPITATION, the collection area for WATER SUPPLIES, or a river system. The area is defined by the notional line, or watershed, on surrounding high land.

Caustic soda (NaOH).   Sodium hydroxide, an alkali used in the manufacture of wood-free PAPER and PULP.

Cellulose.   A carbohydrate polymer formed by the action of PHOTO-SYNTHESIS. It is the chief structural element and major constituent of the cell walls of trees and other higher plants where it is bound up with a glue (lignin) to form plant stems. The fixation of carbon in the carbon cycle is such that about 100 000 million tonnes of cellulose are produced annually. Many important derivatives stem from it, e.g. cellulose nitrate and cellulose acetate.

A major use is in paper manufacture when it has been freed from its matrix of lignin and other organic matter by treatment with sodium sulphite or alkali to yield a pulp feedstock of cellulose fibres.

It is the principal chemical constituent of DOMESTIC REFUSE and can be processed to yield a variety of products such as glucose, SINGLE CELL PROTEIN and ethylalcohol. (⟡HYDROLYSIS; PYROLYSIS; COMPOSTING; RECYCLING.)

Cellulose economy.   The use of the sun's energy through the process of photosynthesis and the subsequent production of CELLULOSE which is then processed, by enzymatic means or by HYDROLYSIS, to yield glucose which can then be fermented to yield many industrial products, protein or fuels.

The cellulose economy would use strains of plants which have a high energy conversion efficiency, e.g. sugar cane and sorghum in tropical zones which can fix 3 per cent of the incoming solar energy. These plants are then processed by the chemical or biochemical means already mentioned. This technique of energy cropping would also use marginal lands, swamps, etc., as productive areas. Species would be used that are capable of continuous cropping, such as fast-growing deciduous trees which resprout from stump when cut and avoid the need for replanting. In the UK and USA, willows and hybrid poplars can be so cropped. One strain of hybrid poplar has been planted at 3700 trees per acre and produces eight harvests per planting with unit energy costs within an order of magnitude of those from fossil fuels.

Cellulose processing as a means of living off the sun's income is a fast-growing area of research and development especially for the production of SINGLE CELL PROTEIN. (⟡ENZYMES.)

Central nervous system, Effects of chlorinated hydrocarbons.   ⟡CHLORI-NATED HYDROCARBONS.

Central nervous system, Effects of DDT. ⇨DDT.

Chelating agents. A group of organic compounds which can incorporate metal ions into their structure and so obtain a soluble, stable and readily excretable substance (known chemically as a complex). Thus toxic metals such as lead (normally poorly excreted) can be removed from blood and bones of the human organism and excreted. The use of chelating agents is a standard treatment in such cases of poisoning.

They are, however, unable to penetrate the body's cellular material and to overcome this one method is to encapsulate them in substances known as liposomes which can penetrate the cell walls. This is an important application as the liver, for example, is one of the vital organs where plutonium and other heavy metals accumulate. Thus intracellular deposits may now be purged from the human organism.

The medical use of chelating agents must be very carefully controlled as the sudden burden of heavy metals released can damage the liver and kidneys.

Another extremely useful application of chelating agents is the use of ethylenediamine-tetracetate (EDTA) to sequester minute quantities of copper, iron, etc., in foods, thereby preventing discoloration. EDTA is also a possible preservative as it can inhibit the bacterial growth responsible for food poisoning.

Chemical oxygen demand (COD). ⇨DISSOLVED OXYGEN; BIO-CHEMICAL OXYGEN DEMAND.

Chlor-alkali process. A process used for the manufacture of CHLORINE by ELECTROLYSIS of brine. This process was formerly a source of mercurial discharge to estuaries which gave rise to MINAMATA DISEASE in Japan.

Chlorinated hydrocarbons. One of the three major groups of synthetic insecticides (the others being organophosphates and synthetic pyrethrins), among which are included DDT, Aldrin, Endrin, benzene hexachloride, Dieldrin, and many others.

In insects and other animals these compounds act primarily on the central nervous system. They become concentrated in the fats of organisms and thus tend to produce fatty infiltration of the heart and fatty degeneration of the liver in vertebrates. In fishes they have the effect of preventing oxygen uptake, causing suffocation. They are also known to slow the rate of photosynthesis in plants.

Their danger to the ecosystem resides in their great stability and the fact that they are broad-spectrum poisons which are very mobile because of their propensity to stick to dust particles and evaporate *with* water into the atmosphere. (⇨PESTICIDES.)

51

Chlorine (Cl). An element of the halogen group. A gas with an irritant smell which has severe effects on the lungs and respiratory system. Produced mainly by the CHLOR-ALKALI PROCESS, chlorine is widely used in the manufacture of organic chemicals, for example, carbon tetrachloride, HALOGENATED FLUOROCARBONS (AEROSOL PRO-PELLANTS), chloroform, PESTICIDES, PVC, and solvents, plastics and bleaching agents.

Chlorine is often used as a disinfectant in water treatment to protect public health (⍚WATER SUPPLY).

Chlorofluoromethanes. A class of chemical compounds commonly used as AEROSOL PROPELLANTS, e.g. TRICHLOROFLUORMETHANE, and refrigerants, e.g. Freon-12. The annual production of these gases is now 1 megaton. Some of it is in captive use, such as refrigeration. The gases are chemically inert, but after absorption of short-wave radiation, each chlorofluoromethane molecule decomposes with the release of the radical atomic chlorine which cannot exist independently and thus attacks OZONE (O_3) through catalytic chain reactions which may lead to ozone depletion.

The gases are characterized by a long atmospheric lifetime because of their chemical and biological inertness, and because their relative insolubility in water prevents rapid removal by cloud formation and precipitation.

The long atmospheric lifetime coupled with the delay period before the chlorofluoromethanes diffuse upwards and encounter the short-wave ultra-violet radiation mean that changes in the ozone layer will be observable through most of the next century even if production were held at current rates. Whether the changes will be of major importance has yet to be determined, but once again the need for environmental acceptability tests for the products of the synthetic chemicals industry is demonstrated.

Chloroprene. A chemical used for the production of synthetic rubber, Neoprene, which has been identified as a possible CARCINOGEN. Workers in a Russian Neoprene factory showed higher than normal levels of lung and skin cancer.

Chlorophenols. Major group of chlorinated hydrocarbon PESTICIDES and biocides which in the UK account for over 85 per cent of the non-agricultural pesticide use, such as anti-rotting agents in non-woollen textiles and wood preservatives.

The chlorophenols act as biocides by inhibiting the respiration and energy conversion processes of the micro-organisms. They are toxic to man above 40 parts per million (ppm), to fish above 1 ppm, whilst concentrations as low as 1 part per thousand million can taint water. Little is known about their ecological effects, which is not to imply that they are harmless.

Chlorophyll. A combination of green and yellow pigments, present in all 'green' plants, which capture light energy and enable the plants to form carbohydrate material from carbon dioxide and water in the process known as PHOTOSYNTHESIS. It is found in all ALGAE, phytoplankton, and almost all higher plants.

Cholera. Bacterial infection spread by contamination of drinking-water supplies by sewage effluent and infected food. It is prevalent in Eastern countries and those areas where the growth in tourism has not been matched by the development of safe water supplies and adequate sewage disposal facilities. Tourist camping sites and recreation centres near lakes and rivers are particularly sensitive areas where precautions must be taken.

Prevention of infection is by the use of separate piped water and sewage systems and chlorination of water supplies before use. (⊘CHLO-RINE; WATER SUPPLY; SEWAGE EFFLUENT TREATMENT.)

Chrome wastes. Chromates and chromic acid are common wastes from certain industries, such as chromium plating and leather tanning. Chromates are soluble in water and are toxic to sewage treatment processes. (The hexavalent ion is the most toxic – it is present in chromic acid – and must be reduced to the trivalent state to form an insoluble product before being released into the environment.)

Chromates act as irritants to eyes, nose and throat, and may cause dermatitis. On prolonged exposure there is liver and/or kidney damage and possible carcinogenic effects.

Recent reports from Japan indicate that over 50 people throughout the country had died of lung cancer due to chrome poisoning, and nearly 500 more were suffering from diseases attributable to the same pollutant. In August 1975 an official survey of five chromium refining plants suggested that 22 deaths had been caused by hexavalent chromium. Although the dangers of hexa-chrome had been pointed out sixteen years ago by the Japanese National Institute of Public Health, it was not until January 1975 that the labour ministry ordered regular health checks on all those who had worked for more than five years in a chromium factory. As a result of these tests, it was found that there had been 30 deaths in Tokyo and 14 in Hokkaido of lung cancer among workers in chromium plating factories. The number of victims discovered is expected to rise considerably as further investigations are carried out.

Hexavalent chromium may also cause abnormalities in chromosomes.

Chromium (Cr). A hard white metal. It is used as a steel-alloying element and in plating. The metal is stable and non-toxic, but there are water-soluble compounds which are extreme irritants and highly toxic. Chromium is a trace element essential for fat and sugar metabolism. Chromium aerosols can affect health in concentrations above 2·5 microgrammes per cubic metre. Its presence as chromates can cause dermatitis at very low concentrations to individuals whose work has sensitized them, by contact with the compounds; for example, cement contains sufficient chromium (0·03 to 7·8 microgrammes per gramme) to cause dermatitis in sensitive people.

See G. L. Waldbott, *Health Effects of Environmental Pollutants*, C. V. Mosby, St Louis, 1973.

Chronic. Medical term used in relation to the effects of certain pollutants, such as asbestos, to describe a response which develops as a result of long and continuous exposure to low concentrations. It is contrasted to 'acute'.

Cities. It is becoming increasingly obvious that many of the ills that afflict twentieth-century man are connected with our industrial processes, which demand that large groups of human beings live in close proximity to one another and to their source of employment. Perhaps the majority of our social ailments are generated by the social, economic and psychological pressures that are inevitable in large cities (◊CROWDING).

It should not be too difficult, bearing in mind individual needs for space, food, water and waste disposal, to estimate upper limits beyond which a conurbation cannot survive. Many existing cities are probably already past the point at which their survival as properly functioning social systems is possible without ever-increasing expenditure of resources which will become more and more difficult to mobilize.

Clarifier. A mechanical device for removing solids from water. It is used for treating the effluent from paper mills to remove fibrous tissues and fillers. It is also used in water treatment for domestic or industrial use. (◊DE-INKING; PULP; PAPER; WATER SUPPLY.)

Climate. The earth's climate has gradually evolved as the atmosphere has stabilized (◊CARBON CYCLE). The controlling factor is latitude, which governs the intensity of incident solar radiation. Atmospheric currents and oceans iron out large inequalities in temperature. Of the incoming solar energy an average 30 per cent is reflected by the atmo-

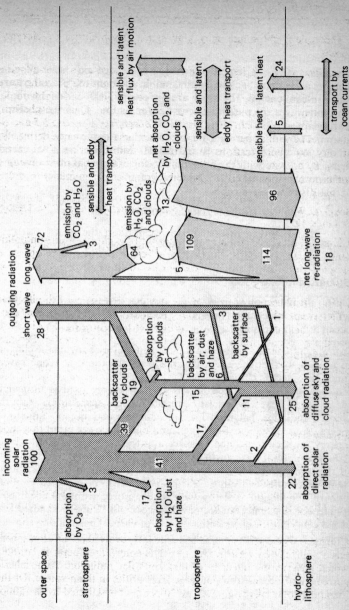

Figure 6. The earth's radiation balance between incoming (solar) radiant energy (on the left) and outgoing (terrestrial) radiant energy (on the right). The figure also shows the distribution of energy in the global system. (Taken from S. H. Schneider and R. D. Dennett, 'Climatic barriers to long-term energy growth', *Ambio*, vol. 4, no. 2, 1975.)

sphere back to space, 50 per cent is absorbed at the earth's surface, and the remainder is absorbed by the atmosphere, which in turn sets up atmospheric circulation patterns. Figure 6 shows on the left the incoming short-wave radiant energy and on the right the outgoing long-wave radiant energy. The input of solar energy (100) equals the output of short-wave (28) plus long-wave (72) radiation. Thus the system is balanced.

A fraction of 1 per cent of the incoming solar energy is fixed by photosynthesis to form trees and plants, and this is the total energy fixation on which life on earth depends. The main man-induced influences on global climatic patterns are:

1. Increased carbon dioxide content (◊GREENHOUSE EFFECT; ENERGY, EFFECT OF CONVERSION ON CLIMATE).

2. A change in the ALBEDO (which can affect the MEAN ANNUAL TEMPERATURE) by clearing jungles, creating reservoirs, and laying vast areas under concrete and asphalt.

3. Dust/aerosol emission, which can reflect ionizing solar radiation.

4. Stratospheric properties may also be affected by SUPERSONIC FLIGHT (◊CONCORDE) or possibly by fluorocarbon propellant dissociation attacking the ozone layer (◊AEROSOL PROPELLANT).

Although these processes have some potential for changing the climate, it is unlikely that there will be drastic modification, but one caveat must be entered. *If the demand for energy continues to grow, the heat balance of the earth could be upset* as all energy used is ultimately degraded to heat which must be reradiated into space to maintain the earth's radiation balance. The introduction and use of 'unnatural' energy sources such as fossil fuels and nuclear power mean that more energy must be reradiated to space and the earth's radiation balance could be upset if the growth continues. The calculations are such that, if the 'unnatural' energy input reaches 1 per cent of the solar input, then major alterations in the earth's climate can be expected as atmospheric circulation patterns will be disturbed, with unknown consequences. Currently we are nowhere near this figure (a hundredth of 1 per cent) but the exponential growth of energy consumption shows it is not entirely inconceivable that this could be reached (◊EXPONENTIAL CURVE). Thus the earth's climate patterns place an ultimate restriction on long-term energy growth. While we are nowhere near the postulated limit, it is important to remember that future energy production may well be in very concentrated locations so that very high energy densities will take place locally and could conceivably cause major regional climate upsets. The atmospheric system is finely tuned and there are limits to what it or any other natural system will tolerate. (◊ THERMAL POLLUTION.)

H. E. Landsberg, 'Man-made climate changes', *Science*, vol. 170, 18 December 1970.
S. H. Schneider and R. D. Dennett, 'Climate barriers to long-term energy growth', *Ambio*, vol. 4, no. 2, 1975.
W. R. Friske, 'Extended industrial revolution and climate change', *American Geophysical Union*, vol. 52, no. 7, July 1971.

Clouds. A mass of water droplets formed in the atmosphere by condensation of water vapour around nuclei such as salt, dust and soil particles. Condensation commonly occurs when there is a drop in air temperature which cools the moist air mass to below its dewpoint (i.e. the temperature at which precipitation occurs in a water-vapour-laden gas stream).

Club of Rome, The. A body set up in 1968 by an international group of economists, scientists, technologists, politicians and others. The Club's object is the study of the interactions of economic, scientific, biological and social components of the present human situation, in the hope of eventually being able to predict, with some degree of certainty, the results of present policies and to formulate alternative policies where it is deemed necessary on environmental and survival grounds. (◊ WORLD MODELS.)

See W. Beckerman, 'Economists, scientists, and environmental catastrophe', an inaugural lecture delivered at University College, London, 24 May 1972.

Coal. The world's coal and lignite (a low-grade coal, often brown in colour, with a relatively low heat value) reserves are greater than those of oil and gas. Two-thirds of these reserves are thought to be in Asia, but they are yet to be proved. Approximately 56 per cent of the world's known coal reserves are in the USSR and Eastern Europe, 9 per cent are in China, 8 per cent in Canada, 5 per cent in Western Europe, 1·4 per cent in Africa, and 0·18 per cent in Central and South America.

World resources of reasonably accessible coal are sufficient to meet projected world demands for the next two to three hundred years, although this estimate would have to be severely reduced if significant quantities are used to make synthetic liquid fuels as a replacement for oil.

The environmental costs of mining coal and lignite are, unfortunately, high. Much of the cheapest can be extracted only by surface mining on a large scale, while the social costs of conventional mining are well known. Furthermore, when used as a fuel, coal is the dirtiest of the fossil fuels. It is possible that enlightened programmes of restoration, of remote mining (see Thring, 1974), and of rigidly enforced controls on air pollution, may minimize such disbenefits to acceptable levels, but the costs involved can be high. (◇TAR SANDS; OIL SHALES.)

M. W. Thring, 'Mole mining', in M. W. Thring and R. J. Crookes (eds.), *Energy and Humanity*, Peter Peregrinus Ltd, 1974, p. 166.

Cobalt–60. Radioactive isotope of cobalt. Cobalt–60 emits high-energy gamma radiation and has a HALF-LIFE of over five years, thus rendering it liable to accumulate in areas of discharge. It is particularly damaging to biological systems because of the radiation hazard, and therefore its discharge is subject to very strict licensing procedures.

Cobalt–60 is an inevitable waste product from nuclear reactors, which use high cobalt steel fuel cans which are, of course, irradiated in the fission process. (◇NUCLEAR REACTOR DESIGNS; IONIZING RADIATION, EFFECTS.)

Coffee, Association with cancer. ◇NITRATE PRESERVATIVES.

Coliform count. A water-purity test: the number of presumptive coliform bacteria present in 100 millilitres of water. The organism *Escherichia coli* is used to indicate the presence of faecal matter which can spread enteric diseases. *Escherichia coli* is the main species of bacteria present in human excreta and while only one in one million of the bacteria may be pathogenic to man, the presence of any *E. coli* is an indication of polluted water. It cannot, therefore, be distributed for consumption without proper sterilization.

Collection tax. This is a form of the polluter-must-pay philosophy. For example, the non-returnable bottle is subsidized by the community who pay for the cost of refuse collection and disposal. A collection tax would make the purveyor/manufacturer of goods in non-returnable bottles pay for their collection and disposal and, of course, may influence an increase in the use of returnable bottles. (◇RECYCLING FINANCIAL INCENTIVES FOR.)

Commons, The. The concept of the commons is one of common ownership of resources of value to the community. Originally the term referred to common land used for pasture, but it has recently been used to describe man's common environmental resources, land, air, water, and so on. The relevance of the analogy has been pointed out by Garrett Hardin, an American biologist, who has shown that the more an individual (or corporation) exploits the commons, the greater the harm to the community as a whole.

G. Hardin, 'The tragedy of the commons', *Science* (NY), vol. 162, 1968, pp. 1243–8.

Composting. A process of controlled decay which enables aerobic bacteria and other micro-organisms to decompose organic matter such as leaves, grasses, paper, and produce a stable end product suitable for soil dressing, or for landfill of DOMESTIC REFUSE in locations where crude refuse cannot be tipped. Sales of compost are governed by the quality, distance to be hauled and state of the market. Compost from domestic refuse is rarely sold as it may be crudely decomposed, contain glass splinters, PLASTICS and HEAVY METALS. (⟡ AEROBIC PROCESS.)

Concentration, Units for. Many concentrations are expressed as parts per million (ppm) or parts per hundred million (pphm), where parts per million is based on proportional parts by volume. However, the International System of Units (SI) uses microgrammes per cubic metre (μg m^{-3}) for air pollution, that is, weight per unit volume, and milligrammes per litre (mg l^{-1}) or grammes per cubic metre (g m^{-3}) for water pollution. The difference in scales for air and water is due to the density difference of a cubic metre of water and a cubic metre of air.

Now in this text, as in many books on the subject, microgrammes per cubic metre values for air pollution are not exclusively used as data are often in parts per million (as in US and many UK tests). To convert from parts per million to microgrammes per cubic metre, we make use of the fact that 1 mole (i.e. a mass of gas equivalent to its molecular weight in grammes) of any gas at standard temperature and pressure occupies a volume of 0·0224 cubic metres (22·4 litres). Thus, the concentration in microgrammes per cubic metre equals

$$\frac{\left(\begin{array}{c}\text{concentration}\\\text{in ppm}\end{array}\right) \times \left(\begin{array}{c}\text{molecular weight}\\\text{of substance}\end{array}\right)}{0 \cdot 0224}$$

For example, sulphur dioxide (SO$_2$) has a molecular weight of

$$\text{S (32)} + \text{O}_2 \text{ (2} \times 16) = 64.$$

Therefore, the concentration in microgrammes per cubic metre of 1 part per million sulphur dioxide is

$$\frac{1 \times 64}{0 \cdot 0224} = 2857 \text{ microgrammes per cubic metre.}$$

Note that for water parts per million by volume is the same as grammes per cubic metre, as 1 gramme of water occupies 1 cubic centimetre, and 1 cubic metre equals 10^6 (one million) cubic centimetres. This relationship applies for dilute aqueous solutions.

Concorde. A supersonic (faster than sound) aircraft which, in order to achieve the very large thrust needed for its flight speed, uses turbo-jet engines which are inherently noisier than the turbo-fan types.

There are two types of noise problem associated with Concorde and other supersonic aircraft: the sonic boom and the extremely large noise 'footprint' (i.e. the area which receives a certain specific noise level). The sonic boom characterizes all supersonic transport (SST) and is produced as the result of a build-up of air pressure in front of the aircraft. From American studies, the pattern of the Concorde 100 EPNdB* footprint ranges 54 square miles, some 41 times larger than the DC–10–30 footprint. Any substantial reduction will require major engine redesign. Allegations have already been made that the British government has concealed the full extent of the noise problem by not constructing a noise footprint for Concorde flights from Heathrow and that they have failed to measure noise levels in residential areas close to the runway.

To date Concorde has broken Heathrow noise rules (measured in PNdB) on 26 take-offs out of 37, and has repeatedly created noise levels above the threshold of pain (133 decibels) (*Observer*, 19 October 1975). Also, the low-frequency sound emitted by Concorde is five times as intense as that from a 707. This type of sound can cause structural damage to buildings.

In addition to noise problems, there are fears that the exhaust emissions in the stratosphere could have far-reaching and destructive effects on climatic conditions. The stratospheric ozone layer may be decreased, causing an increase in the amount of ultra-violet solar radiation reaching the earth's surface. This aspect has yet to be proven and it is argued by the makers that the very much greater number of supersonic flights undertaken by military aircraft do not appear to have damaged the stratospheric ozone layer.

See 'Concorde and the environment', *Flight International*, 27 November 1975, pp. 779–82.

Conservationists. For many years, conservation groups have been engaged in an often bitter struggle to preserve unique areas of the earth and save various endangered species from extinction. Unfortunately, the very nature of such groups is essentially defensive and the battle being fought is a losing one, because it is inevitable that any gains are purely temporary, whereas the losses are permanent. Against powerful

---

* EPNdB – effective perceived noise level: unit used in noise certification based upon perceived noise decibels (PNdB) corrected for particular pure-tone characteristics of jet noise that produce extra annoyance over and above that measured by PNdB. (◊DECIBEL; NOISE CONTROL; PNdB.)

economic and political interests, which can always offer increased employment and higher wages as a result of their development plans, the conservation groups can only resist by a planned rear-guard action in which compromise is usually inevitable. Compromise usually means abandoning some part of our natural heritage to the demands of 'progress'. Nevertheless, the recent growth of interest in and awareness of environmental problems is largely due to these dedicated groups. (◊COST–BENEFIT ANALYSIS.)

Conservation of land.   The maintenance of areas of the countryside for leisure and the preservation of threatened species of animals and plants is one of the main objectives of a conservation policy. The existence of two bodies in the United Kingdom, the Nature Conservancy and the National Trust, indicates the British commitment to conservation. Conservation is essentially the preservation of man's environment in a condition to fulfil his needs for a healthy and satisfactory life as the pace of living and pressures of industrialized society increase. This is one aspect of the COMMONS that requires eternal vigilance.

Consumers.   ◊FOOD CHAIN; ECONOMIC GROWTH.

Consumers' sovereignty.   ◊ECONOMIC GROWTH.

Continuous sampling.   Uninterrupted sampling of, say, air, usually at a fixed rate. Where the sample can be analysed continuously, the stream of gas may be passed through the measuring instrument continuously. Otherwise the sample is collected in an uninterrupted fashion for a given period and the total sample is finally analysed to give the mean composition of the air over the whole period.

Contraception.   BIRTH CONTROL by the use of contraceptive devices. The most common of these are the condom, the diaphragm, the cervical cap and the intrauterine device (IUD). In recent years the contraceptive pill has also become commonplace, as have sterilization and vasectomy techniques (although these methods cannot be classified as contraception). (◊ABORTION.)

Controlled tipping.   The most common method of DOMESTIC REFUSE landfill disposal where the refuse is tipped in layers, compacted and covered at the end of every working day with an inert layer of suitable material to form a seal. Controlled tipping avoids refuse being blown from the site, or tipping in static or running water; and ensures the general orderly appearance and running of the tip. The opposite of this practice are open tips which constitute public nuisances and can threaten public health by providing breeding grounds for rodents and flies.

Copolymerization. ▷POLYMERIZATION.

Copper (Cu). A metal, commonly used for heat-transfer applications because of its high thermal conductivity. It is a very good electrical conductor and can be readily drawn or extruded into tubes and wire.

Copper is an essential constituent of living systems. However, copper ions (Cu^{++}) are toxic to most forms of life, 0·5 parts per million being lethal to many algae. Most fish succumb to a few parts per million. In higher animals, brain damage is a characteristic feature of copper poisoning.

Copper pollution may arise from many sources. Soils receive high levels as a result of mining activities, intensive use of 'copper pellets' in pig rearing, or the application of copper fungicide. Effluents from factories and mines can cause serious water pollution. (▷MATERIALS RESOURCES.)

Corrected noise level (CNL). Index of industrial noise. The level of noise emitted from industrial premises in dB(A), corrected for tonal character, intermittency and duration in accordance with the appropriate British Standard. The CNL measure is required because of the intermittent nature of many industrial noise sources. (▷DECIBEL A-SCALE; INDUSTRIAL NOISE MEASUREMENT.)

Cost–benefit analysis. A technique which purports to evaluate the social costs and social benefits of investment projects in order to help decide whether or not such projects should be undertaken.

Cost–benefit methods have been offered to support the nuclear industries' claim that zero radiation release is too expensive. By using cost–benefit analysis, the minimum radiation levels that society can 'afford' to tolerate may be calculated. This approach is embodied in Figure 7, the thesis being that the dose that is 'as low as is readily achievable, economic considerations being taken into account', is thereby obtained.

One major objection to this approach is that exposure to radiation is an involuntary risk and is not one that an individual can opt out of. For example, when a person uses his car he takes voluntary risk of an accident happening as opposed to completing his journey (benefit) and the costs and benefits apply to the same individual. In the case of nuclear power, costs and benefits do not apply to the same individuals and risks cannot be eliminated by voluntary action.

The questions which cost–benefit analyses have failed to answer in this and other areas are:

1. How do you put a financial cost on a risk when there is no mechanism for people to buy themselves out of the risk?

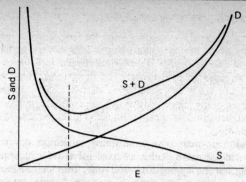

Figure 7. Differential cost–benefit analysis reveals the radiation level from nuclear operations that is, according to the International Commission on Radiological Protection, 'as low as is readily achievable, economic and social considerations being taken into account.' The two curves S and D represent the full cost of achieving a standard of safety resulting in the collective dose (E), and the cost of the detriment resulting from that dose. Combining these two curves gives the U-shaped curve representing the total cost of the collective dose. The minimum of the S+D curve is the dose that is 'as low as readily achievable'.

2. How can 'emotive issues' be taken out of cost–benefit analysis when ultimately all value is derived from emotion?

3. How can one evaluate either costs or benefits to non-direct participants in a system which does not yet exist?

4. On what evidence is the same discount rate applied to costs and benefits? In particular, why apply a discount rate to children? Most people seem to apply an interest rate to their children.

Among economists themselves, perhaps the most devastating indictment of cost–benefit analysis has come from Schumacher:

It can therefore never serve to clarify the situation and lead to an enlightened decision. All it can do is lead to self-deception or the deception of others; for to undertake to measure the immeasurable is absurd and constitutes but an elaborate method of moving from preconceived notions to foregone conclusions; all one has to do to obtain the desired results is to impute suitable values to the immeasurable costs and benefits. The logical absurdity, however, is not the greatest fault of the undertaking: what is worse, and destructive of civilisation, is the pretence that everything has a price or, in other words, that money is the highest of all values (E. F. Schumacher, *Small is Beautiful*, Blond & Briggs, 1973).

63

The controversy over cost–benefit analysis is central to the study of the environment.

For a full discussion of the issues involved, Stanislav Andreski's *Social Sciences as Sorcery*, Penguin, 1974, is recommended.
The correspondence following J. Dunster's article, 'Costs and benefits of nuclear power', in the *New Scientist* of 18 October 1973 (*New Scientist*, 1 November, 8 November, 22 November, 29 November) includes important points in the debate over the application of cost–benefit analysis to nuclear energy.

Critical group. A method of monitoring pollution emission which relies on the identification of the group of individuals most at risk to a particular discharge. If they are receiving doses or exposures below those recommended, it is assumed that the population at large is therefore not exposed or subjected to the same discharge. This technique has been used for the monitoring of IONIZING RADIATION emissions from nuclear reactors. It depends on the success with which the group most at risk can be identified.

It is also proposed as a method of controlling discharges of HEAVY METALS to estuaries. For example, crab meat from South Devon contains up to 21 parts per million of CADMIUM, and if the group of individuals were identified as, say, 'large' crab-meat eaters, and if their daily intake was below the Food and Agricultural Organization/World Health Organization limit of 60 microgrammes for a 70-kilogramme man, then the population at large could be assumed to be not at risk.

Children and old people are two distinct critical groups that require separate consideration. Children take in approximately twice as much food as an adult in relation to body weight. Thus their exposure to risk is greater. Old people generally have weaker internal organs, and this again may militate against bodily defence mechanisms functioning correctly.

Critical pathway. In planning a nuclear installation, the environmental pathway is found for every nuclide that is to be released on discharge through food chains or otherwise to any member of the population. Some member(s) of the population is thereby found to constitute the critical case of exposure, i.e. CRITICAL GROUP. If it is reasonably certain that the dose in the critical case is less than the International Commission for Radiological Protection (ICRP) limit (⟡IONIZING RADIATION, MAXIMUM PERMISSIBLE DOSE), then no other persons will be at any greater risk. However, ICRP recommends that every exposure should be reduced as low as is readily achievable within the economic and social framework. Efforts are also made to exploit any opportunity to reduce the dose still further.

As an example, the critical exposure for Windscale is to an eater of edible seaweed (from which a delicacy called laver-bread is made). The isotope producing the dose is ruthenium–106, and in 1970 the estimate of the dose, assuming that all the seaweed is contaminated to the greatest extent measured and that an improbably large amount is consumed regularly, was 6 per cent of the I C R P's recommended limit. The monitoring is done by the Fisheries Radiobiological Laboratory, and a watch is maintained on other isotopes which include zirconium–95, niobium–95 and caesium–137. For the first two of these, the critical group is fishermen who are exposed to external gamma radiation in their contact with estuarine silt, and for caesium–137 the critical group is anglers. In 1970 the fishermen received 10 per cent of the limiting dose and the anglers 3 per cent. This example illustrates the care taken in the U K in assessing the risks to the population in the vicinity of a nuclear installation.

Crops, New strains. ⟡ GREEN REVOLUTION.

Crowding. Experiments carried out on rats kept under conditions of severe overcrowding have revealed an apparent breakdown in the established social system, with a marked increase in aberrant behaviour, including neglect of offspring, cannibalism and violent aggression. In addition, the incidence of miscarriages and sterility increases and the death rate rises.

Autopsies conducted on the rats indicate stress-induced exhaustion of the adrenal cortex. Although there is as yet no direct medical evidence of stress-induced diseases, the prevalence of suicides, child abuse, mental illness and violent crime in our larger cities makes the temptation to draw parallels almost irresistible. Such temptations should be resisted. First, the sociological evidence is conflicting, and some studies tend to indicate that there is little or no increase in mental illness with urbanization. Second, aberrant behaviour and mental illness in human society are particularly hard to define. If one defines mental illness merely as behaviour generally considered disturbed by the majority of society, then this can be the thin edge of a very dangerous wedge – as the treatment of political dissidents in some countries indicates.

Human beings do not respond to crowding in the same way as rats, which have a very elementary social structure and almost no flexibility in their behaviour. Any minor modification to the environment can thus destroy the existing social structure. Human beings, on the other hand, are extremely adaptable and gregarious, and have a complex social structure capable of great flexibility.

Undoubtedly cultural differences are an important factor – Japan, Holland and the UK, while being vastly more crowded than the USA, have a much lower crime rate. ($\diamond$ CITIES.)

J. L. Freedman, *Crowding and Behavior: the Psychology of High-density Living*, Viking Press, 1975.

Cryogenic scrap recovery. A form of scrap recovery which uses liquid nitrogen to freeze (at $-196°C$) the pot-pourri of metals and contaminants from used cars. The very low temperature causes mild steel to fracture like glass when put through an impactor. The steel can then be easily removed from its trapped contaminants magnetically and the result is a very high-grade scrap suitable for high-quality steel manufacture. The contaminants, which include copper and zinc, may be further processed for high-value scrap recovery. This method of recovery is being tested on tin cans in DOMESTIC REFUSE in north-east England. It allows excellent metals separation and 6000 tonnes per year of scrap steel are currently being recovered.

Cullet. Trade name for colour-graded glass fragments, offcuts, etc., suitable for remelting. Sources are mainly the glassworks themselves and bottling plants. DOMESTIC REFUSE is another source but the RECYCLING cost is often prohibitive when compared with the cost of the raw materials themselves as collection and transport costs outweigh the revenue from the sales.

curie (Ci). Unit of radioactivity defined as 37000 million (3.7×10^{10}) disintegrations per second which is approximately equal to the activity of 1 gramme of radium. Thus an amount of radioactivity is expressed effectively by stating the number of grammes of radium that could provide the number of disintegrations per second. In practice it is far too large and the picocurie is used, i.e. 1 million millionth of a curie or 10^{-12}.

Cycle. A process or series of operations performed by a system in which conditions at the end of the process are the same as the original state. Thus in the CARBON CYCLE the carbon dioxide in the air is fixed by photosynthesis into plant life which in turn decomposes or is consumed and eventually ends up as carbon dioxide again.

Cyclone dust separator. A device for removing dust particles from air. The principle is shown in Figure 8. The dust-laden gas enters tangentially and is spun in a helical path down the conical collector. The particles are flung to the wall by centrifugal force where they drop into the dust hopper while the clean gas leaves through a central 'core' tube at the top. Cyclones can remove 75–95 per cent of particulates in their range of applicability, for smaller particle sizes ELECTROSTATIC PRECIPITATORS are often used, e.g. to remove FLY ASH.

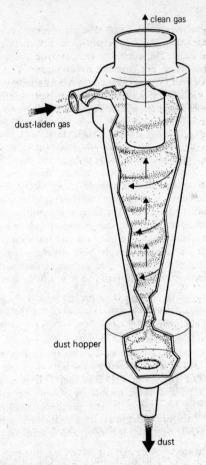

Figure 8.   Cyclonic dust separation.

D

Dam projects. Recent major dam projects have provided examples of high technology applied with the most benevolent of motives but with inadequate consideration of any possible ecological side-effects. The most obvious example is the Aswan Dam. Designed to produce power and to store water for a permanent irrigation system, the effect of the irrigation ditches on the spread of a serious disease, SCHISTOSOMIASIS (Bilharzia) was not considered. Similar unexpected side-effects occurred with the Kariba Dam, which helped spread a fly-borne disease which has disrupted the agriculture of people living along the river banks. Furthermore, for most water projects of this type, silting of the reservoir will eventually eliminate all the temporary gains.

In the case of the Aswan Dam, the changes produced in the flow of the Nile have had detrimental effects on the fisheries of the eastern Mediterranean. It will also probably have a deleterious effect on the soil fertility in the Nile Delta, since nutrients that were previously deposited annually by the flooding of the Nile will now be absent.

An admirable survey of the Aswan Dam project is given by Rex Keating in the Open University 'Environment File', Unit 23 of the Technology Foundation Course 'The man-made world', *The Aswan High Dam and its Effects on the Environment*, Open University Press, 1972.

DDE; chemical name: 1, 1-di-chloro-2. 2-bis (p-chlorophenyl) ethylene. A metabolite of DDT, it is formed by the action of some soil micro-organisms on DDT (in water DDT is inert). It is more inert and persistent than DDT and large quantities of both products could be accumulating in the biosphere. In the presence of ultra-violet light (i.e. sunlight) it can photodegrade to about nine other products, including several chlorinated biphenyls which can also have severe biological effects.

DDE is one of the most abundant organochlorine compounds in the biosphere.

DDT (Dichlorodiphenyltrichloroethane). An organochlorine PESTICIDE developed during the Second World War as a delousing agent and later

68

used to combat the insect carriers of malaria, yellow fever and typhus, thereby saving many lives. However, in the early 1960s certain immune strains of mosquitoes and other disease carriers developed. At the same time a decline in the reproduction rate of fish and bird life has occurred.

The key problems are the persistence of DDT and its high biological activity. It is estimated that there are several million tonnes in circulation in the biosphere, spread by air and water currents, and most of this ends up in the ocean. Small concentrations (0·01 part per million) reduce photosynthesis in marine plankton by 20 per cent, and 1 part in 1 thousand million (10^9) parts sea water killed 39 per cent of brine shrimps in three weeks (see Wurster, 1968).

Ecological effects show that often the pests have greater powers of recovery than their predators or parasites, so the use of DDT can, in some cases, be counter-productive – a classic example is the red spider mite. There is little evidence that DDT has adversely affected man as yet, but it undoubtedly has affected wildlife considerably.

C. F. Wurster, *Science*, vol. 159, 1968, p. 1474.

DDVP (Dichlorvos-2, 2-dichlorovinyl dimethyl phosphate). An insecticide often used in pest strips sold for domestic use. The Environmental Protection Agency has cautioned against its use in rooms where food is prepared, or where infants or aged persons are confined – for example, in hospitals. Dietary deficiencies of protein, minerals, or vitamins may increase any ill effects. (◊ PESTICIDES; ORGANOCHLORINES.)

Decibel (dB). A logarithmic measure used to compare the sound level of interest with a reference level. If we are concerned with sound power then reference is made to the smallest sound power that can be heard by someone with normal hearing at 100 Hz. This reference power is 10^{-12} W. As an example we can deduce the sound power level of a jet aircraft on take-off. Now the noise of a jet aircraft at take-off (100 metres) is approximately 1 W which is 10^{12} as powerful as the reference power. Therefore, it is said to differ from the reference sound by $\log\left(\dfrac{\text{power}_2}{\text{power}_1}\right) = \log 10^{12} = 12$ bels. But bels are too large for convenience and so decibels are used instead, i.e. a factor of 10 is introduced. Hence the jet aircraft at take-off has a sound power level of 120 dB with reference to a power of 10^{-12} W.

Sometimes it is necessary to compare sound pressures. Power is proportional to the mean square pressure under reflection-free conditions $dB = 10 \log\left(\dfrac{\text{power}_2}{\text{power}_1}\right) = 20 \log\left(\dfrac{\text{pressure}_2}{\text{pressure}_1}\right)$. Thus, an increase of 3dB in the sound power level corresponds to a doubling of the sound power which corresponds to an increase of 6dB in the sound *pressure* level.

The specification of the scale is important. For most purposes loudness is quoted in decibels A-scale (dB(A)) although other scales are used as well (◊HEARING; SOUND).

Decibels are also used in telecommunications work as a measure of the system response (e.g. signal-to-noise ratio).

Decibels A-scale (dB(A)). A frequency weighted noise unit widely used for traffic and industrial noise measurement. The decibels A-scale corresponds approximately to the frequency response of the ear and thus correlates well with loudness. Other noise scales are used as well but the common ones are dB(A) and PNdB. (◊DECIBEL; SOUND; HEARING; PNdB.)

Decomposers. Organisms, usually BACTERIA or FUNGI, which use dead plants or animals as sources of food. They break down this material, obtaining the energy needed for life and releasing minerals and nutrients back into the environment where they are assimilated by plant and animal life. They are an essential part of natural cycles and enable the components for life to be recycled. (◊CARBON CYCLE.)

Defoliants. ◊HERBICIDES.

De-inking. A process for removing ink from printed paper used in RECYCLING. De-inking takes place in four stages:

1. Pulping of the reclaimed paper with soda or other chemicals.

2. Centrifugal cleaning to remove paper clips, staples and dirt.

3. Screening to remove the freed fibres.

4. Washing to remove the printing ink solids either by mechanical or flotation methods.

The yield is 65–75 per cent of the input paper.

PULP recovered from paper made of groundwood can be used for newsprint, magazines, and as a base for coated papers. Pulp recovered from paper free from groundwood can be used for the manufacture of writing, printing papers and tissues.

The effluent from de-inking plants has a high BIOCHEMICAL OXYGEN DEMAND and requires considerable treatment before discharge.

Demand. The amount of a commodity that people are prepared to buy. If the price of a commodity varies, it would be reasonable to suppose that, at higher prices, less of the commodity will be bought.

The fact that people still want the commodity does not mean that they demand it. If they cannot afford it, they exert no demand for it.

A graph showing the quantities of a commodity demanded at various prices is called a demand curve (see Figure 9). This shows, for each possible price, the total intended purchases of all buyers.

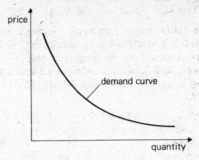

Figure 9. The demand curve.

Denitrifying bacteria. ▷NITROGEN CYCLE.

DES (Diethylstilboestrol). A growth promoting hormone used in the USA in the cattle industry to cut down on feed required to produce a given weight gain. It has been found to be CARCINOGENIC in animals and is therefore a cause for concern in connection with humans.

Desalination. The partial or complete removal of the dissolved solids in sea or saline water to make it suitable for domestic, agricultural or industrial purposes. The main techniques are electrodialysis, reverse osmosis, freezing, and distillation. The processes and their applications are shown in Figure 10.

Electrodialysis
An electric current is passed through brackish or low salinity water in a chamber in which many closely spaced ion-selective membranes are placed, thus dividing the chamber into compartments. The electric current causes the salts to be concentrated in alternate compartments, with reduced salt content in the remainder, thus producing a reduced salt product. A principal disadvantage of electrodialysis is that power consumption is proportional to total dissolved solids or salinity.

Reverse osmosis
The reverse application of osmotic pressure. When salt water and fresh water are separated by a semi-permeable membrane, osmotic

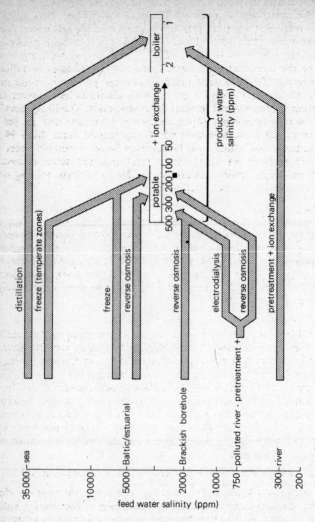

Figure 10. Desalination process application.

pressure causes the fresh water to flow through the membrane to dilute the saline water until osmotic equilibrium is established. Now applying this in reverse, if a greater pressure is applied to the salt water side of the membrane, then relatively pure water will pass through it leaving a concentrated brine to be disposed of. The principal application may well be the treatment of industrial and sewage effluents or polluted rivers in the 750–1000 parts per million range either for re-use or to obtain an acceptable discharge. As with electrodialysis, power consumption is proportional to total dissolved solids. The likely development is a progression towards sea water treatment but as yet no membrane has been developed which can sustain better than 99 per cent salt rejection which must be achieved before a potable water can be obtained from sea water with a total dissolved solids content of 35 000 ppm. The membranes are susceptible to bacterial fouling when applied to certain classes of effluent.

Freezing

The freezing of a salt solution causes crystals of pure water to nucleate and grow, leaving a brine concentrate behind. One commonly proposed freezing method is the use of a secondary refrigerant in which butane is evaporated in direct contact with sea water, resulting in the formation of ice crystals. The ice is separated, melted by the compressed butane vapour, the two liquids are decanted and the product water obtained and the butane recycled.

Distillation

The boiling or evaporation of sea water to form water vapour which is then condensed to yield a salt-free stream. Energy requirements are virtually independent of the feed-water salinity, and product purity of less than 50 parts per million can readily be achieved.

There are two main classes of distillation: heat consuming and power (mechanical energy) consuming. A heat-consuming process has lower energy input costs compared with the power-consuming process, as dictated by the conversion of heat to power. Over 85 per cent of the world's installed desalting capacity is accomplished by heat-consuming distillation processes.

See A. Porteous, *Saline Water Distillation Processes*, Longmans, 1975.

Destructive distillation. The heating of solid substances in closed retorts in the absence of air, and condensation of the ensuing volatile gases. The process has been mooted for the utilization of DOMESTIC REFUSE for the production of METHANOL. It is akin to PYROLYSIS which is often used synonymously for the destructive distillation of refuse.

Detergents. Cleaning agents which include, as part of their chemical make-up, petrochemical or other synthetically derived wetting agents. They are made up of three main parts:

1. Surfactant – a wetting agent which permits water to penetrate fabric more. The surfactant molecules provide a link between the dirt molecules and the water molecules.

2. Builder – a sequestering agent. These tie up hard water ions to form large water-soluble ions. The water becomes alkaline, which is necessary for removal of dirt.

3. Miscellaneous – brighteners, perfumes, anti-redeposition agents and enzymes.

A major drawback in the use of 'hard' detergents is that they are non-biodegradable. There is now a move to incorporate only bio-degradable substances in detergents – although the use of hard detergents is considered necessary in certain industries such as wool scouring. The decomposition or breakdown of detergents leads to phosphorus becoming available in aquatic systems and where this is the limiting nutrient it can cause EUTROPHICATION. (◊DETERGENTS, SUGAR-BASED.)

Detergents, Sugar-based. Tate and Lyle have recently been testing a new detergent that is made by reacting sugar directly with tallow – a product made from animal fat. The result is a detergent which is completely biodegradable, contains no phosphates, and is claimed to be just as effective as conventional products.

Toxicity tests indicate that the product is quite harmless. It is hoped that the product will be the forerunner of a whole new chemical industry based on sugar (a renewable resource) rather than oil or coal.

Developing nations. ◊UNDERDEVELOPED COUNTRIES.

Dieldrin. ◊CHLORINATED HYDROCARBONS; ALDRIN.

Dioxin. A family of chemicals, the major member of which is 2,4,7,8-tetrachlorodibenzo-p-dioxin (TCDDD), which is a manufacturing impurity in certain classes of HERBICIDES and has caused still-births (TERATOGEN).

The US government has a military defoliant, 'Agent Orange', and caused a major row when it planned to sell its surplus stockpile (or part of it) – approximately 2 million gallons – to South America. Agent Orange contains 28 times the maximum acceptable safety limit of dioxin, which is 0·1 part per million for new herbicides. Dioxin is environmentally stable and can thus enter food chains.

Direct energy conversion. The conversion of chemical, solar or nuclear energy directly into electricity without producing mechanical work in the process (as in the conventional boiler–steam turbine–generator system). Direct energy conversion is a very desirable goal since it means that electricity could be made without intermediate equipment and perhaps with greater efficiency.

The classes of direct energy conversion devices are: FUEL CELLS, magnetohydrodynamic generators, THERMIONIC CONVERTERS, and semiconductor THERMOELECTRIC CONVERTERS.

The first law of thermodynamics applies to all the above devices as does the second law in its general form, but the CARNOT EFFICIENCY restrictions do not apply to all direct energy converters. Thus, in theory, direct energy conversion could offer very attractive conversion efficiencies when practised.

Discounted cash flow. A common economic technique used to evaluate the relative costs of proposed schemes.

Knowing the rate of interest available, we can calculate what value any given sum will have at some future date. Conversely, if we know that, at some time in the future, we are going to have to spend a given sum, then we can easily work backwards to determine the present sum that must be put aside for such a future project. This process is called 'discounting'.

Interest rates give us the amount by which an investment will grow annually. In discounting, the term 'discount rate' is used to signify the annual rate of interest at which the present value of the required future sum would have to grow over the intervening period in order to become that future sum.

In appraising projects, the term 'rate of return' is also used to indicate the return on investment that the project will generate.

The construction of any project involves the payment of cash at intervals, and its subsequent operation invariably involves further periodic payments (salaries, material, maintenance, etc.). Expressed on an annual basis, such a stream of payments represents a cash flow (outwards in this case, but inwards if you are calculating benefits, e.g. payments for goods or services resulting from the project).

Now just as we can calculate the present value of a single future sum, so also can we obtain the present value of a series of future sums (costs or benefits), and such a procedure is called discounted cash flow. It is used to evaluate, on purely monetary grounds, the relative merits of various alternative projects or alternative ways of implementing a single project. A common application is in least-cost analysis. For example, in laying on a needed water supply, do we use a large pipe and let the water flow by gravity (high initial capital investment, low running costs) or a smaller pipe in association with pumps (lower capital investment, higher running costs)? Using discounted cash flow methods, we can calculate which of the two schemes will cost least in terms of the present value of the sums to be committed. (It is customary, in such calculations, to ignore costs that will be common to both schemes, such as pipeline maintenance, repairs, metering, etc.)

Dissolved oxygen (DO).   The amount of oxygen dissolved in a stream, river or lake is an indication of the degree of health of the stream and its ability to support a balanced aquatic ecosystem. The maximum amount of oxygen that can be held in solution in a stream is termed the saturation concentration and, as it is a function of temperature, the greater the temperature, the less the saturation amount. It is customary to express oxygen concentrations as percentages of saturation. The presence of dissolved salts can also reduce the saturation concentration at any given temperature, thus sea water contains less oxygen than fresh water at the same temperature and saturation conditions.

The discharge of an organic waste to a stream, e.g. sewage, imposes an oxygen demand on the stream. If there is an excessive amount of organic matter, the oxidation of the waste will consume oxygen more rapidly than it can be replenished. When this happens, the dissolved oxygen is depleted and results in the death of the higher forms of life. In extreme cases the river becomes anaerobic ($\diamond$ANAEROBIC PROCESS), the oxidation processes cease and noxious odours are given off.

The oxygen demand of the waste is of crucial importance and two measures are commonly made.

1. BIOCHEMICAL OXYGEN DEMAND (BOD) is a measure of the oxygen required by the microbes which reduce the wastes to simple compounds, as in SEWAGE EFFLUENT TREATMENT. The BOD is a standard test of the amount of oxygen required by a sample of effluent over five days at $20°C$ and is stated as the parts per million of oxygen (milligrammes per litre) taken up by the sample of effluent incubated in the dark.

2. Chemical oxygen demand (COD) measures the number of parts per million of oxygen taken up by a sample from a solution of boiling potassium dichromate in two hours. The BOD and COD tests differentiate between materials that can be oxidized biologically and those that cannot, and indicate what types of treatment will be required.

It should be pointed out that the above criteria are not in themselves adequate indicators of pollution. There are extremely toxic solutions of cyanide, for example, that have acceptable BOD and COD values.

The oxygen balance of a river is of great importance in controlling the effects of pollution. As soon as organic matter enters the water it exerts a BOD, the dissolved oxygen level falls and as a result an oxygen deficit is created, which means that oxygen transfer from the air to the water will take place to try to restore saturation. The rate of oxygen transfer, i.e. the re-aeration rate, depends on the nature of the flow, the stream-bed characteristics, and the water temperature. A mountain stream will have a re-aeration rate many times greater than a stagnant quarry pond, and would thus be able to assimilate a much greater BOD load per unit volume without a serious oxygen deficit.

As the amount of oxygen consumed by the water increases, the oxygen deficit also increases and in turn the re-aeration rate increases until (if the pollution does not swamp the watercourse oxygen completely) the rates of deoxygenation and re-aeration eventually balance. From then on the re-aeration supplies more oxygen than is consumed by the BOD and the stream eventually recovers to saturation.

This process can be represented by an oxygen sag curve, which is a graph of dissolved oxygen against time (Figure 11). With moderate

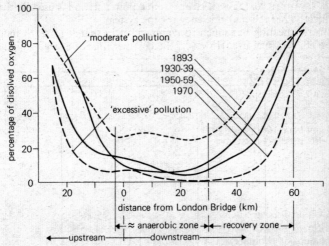

Figure 11. Oxygen sag curves for the river Thames. (From the Royal Commission on Environmental Protection, *Report No. 1*, HMSO, 1971.)

pollution the dissolved oxygen levels are high enough to support fish life, while with high pollution the stream becomes anaerobic.

Distillation. A process by which mixtures of liquids are heated and evaporated. Each vapour then condenses back to a liquid at a characteristic temperature and can be separated. Dissolved solids are left behind.

Distillation is a principal DESALINATION process for the production of pure water from saline water. Note that distillation is a two-stage process: first, evaporation and then condensation of the vapour. In the first stage the latent heat of evaporation has to be supplied to the liquid, and in the second stage the latent heat of condensation (virtually identical to the latent heat of evaporation) has to be removed from the vapour for it to condense.

District heating. The use of a large, efficient, centralized boiler plant or 'waste' steam from a power station (⬦ENERGY) to heat a district and/or supply steam to small industries. The heat is distributed by means of low-pressure steam or high-temperature water mains to the consumers.

As the boiler plant is centralized, it combines greater utilization of fuel compared with a myriad of small domestic or industrial appliances, together with a considerable reduction in low-level pollution compared with, say, open coal fires. If steam from a power station is used, it gives a much higher thermal efficiency for the system.

District heating has much to commend it in that it leads to much more efficient fuel use. However, the cost per unit of energy supplied

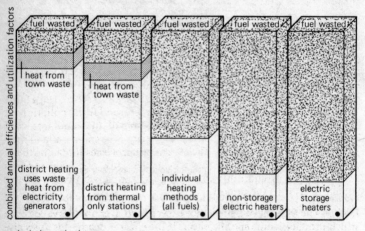

Figure 12. Primary fuel utilization factors for alternative heating methods. (From *Coal and Energy Quarterly*, no. 7, Winter 1975.)

must be competitive with that of alternative fuels. This requirement means that the customers must live very close to the boiler plant or power station, otherwise the cost of distribution makes the energy selling price prohibitively expensive.

DNA (deoxyribonucleic acid). An essential compound found in the nucleus of living cells. DNA is a long-chain molecule, which contains the genetic codes necessary for the development and functioning of the human organism. It controls the formation of PROTEIN and ENZYMES. Carbon atoms make up 37 per cent of DNA, and in exposures to IONIZING RADIATION it is entirely possible that a radioactive isotope known as carbon–14 may become incorporated in the DNA structure, thus causing a random change which may have genetic consequences.

Small changes in the complex DNA structure can lead to genetic disorders such as haemophilia, or mental and physical deformity.

Domestic refuse (UK); **Garbage** (USA). The generic name for waste emanating from households. In the UK it has an average percentage composition of

paper and carboard	38	dust and ashes	17
vegetable and foodstuffs	20	glass	10
plastics	1–2	metals	10
rags	2	unclassified	2

In other words, it has approximately 60 per cent organics and 40 per cent inorganics. The organic portion is suitable for PYROLYSIS, HYDROLYSIS, COMPOSTING, or INCINERATION. The inorganics are sources of metals and glass CULLETT, while the ash and cinders may also be combustible. The UK (1974) production was 18 million tonnes or approximately 1 tonne per day per 1000 population.

Most domestic refuse is still disposed of by tipping where steps must be taken to preserve public health by the prevention of pollution of groundwater and blown refuse, the breeding of flies and rats, and unsightly heaps or offensive odours. Tip space is running out and incineration is replacing tipping. The alternative processes listed above are also applicable, especially as they recycle the refuse and make use of a resource which would otherwise be wasted and which many consider should be used to the utmost.

Dose. The amount of a substance experienced over an interval of time which produces some specific effect. If the effect is death, this gives rise to the expression LD_{50} (lethal dose), which is the dose large enough to kill 50 per cent of a sample of animals under test. (◊HALF-LIFE; MERCURY; LETHAL CONCENTRATION.)

Dry weight. The basis used for expressing the concentration of a compound in living organisms. For example, fish have a moisture content of 80 per cent, so that a concentration of 5 parts per million dry weight would only be 1 part per million wet weight in fish.

Dust; often called airborne particulate matter. Solids suspended in air as a result of the disintegration of matter. There is a wide range of particles from 0·0002 microns to 500 microns in diameter. Particles above 10 microns in diameter are not carried far from their source except by strong winds.

The presence of dust particles in the ATMOSPHERE could cause either a net cooling or a net warming, depending on the properties of the particles and the underlying surface. It is commonly supposed that the presence of atmospheric dust will cause the earth to cool, but as well as reflecting radiation from the sun, they will also reflect heat released from earth. Thus particles with a 'white' or 'grey' upper surface and 'black' lower surface would cause cooling of the earth; if they were reversed, warming of the earth would result.

Calculations show that the aerosol extinction coefficient – a measure of how much heat the particles absorb – has a critical value at which there is heating or cooling of the earth's surface. There is a balance in temperature on surfaces that have ALBEDOS in the range 0·35 to 0·60. For albedos greater than 0·6 there is heating, less than 0·35 cooling.

Atmospheric science has a long way to go, but the use of the atmosphere as a sink for pollutants should clearly not be encouraged.

See R. A. Reck, 'Aerosols in the atmosphere. Calculation of the critical absorption/backscatter ratio', *Science*, vol. 186, no. 4168, 13 December 1974, pp. 1034–6.

E

Earth Resources Technology Satellites (ERTS). Orbiting laboratories launched by the National Aeronautics and Space Administration of the USA to monitor natural resources for man. The ERTS are designed to provide systematic repetitive global land coverage. They complete fourteen orbits per day photographing three strips 185 kilometres wide in North America, and eleven strips in the rest of the world.

The orbit has been designed in conjunction with the photographic settings to pass over any location on the earth's surface once every eighteen days. Thus, the repetition allows changes in the earth's surface features to be monitored over time. The possibilities for monitoring ocean dumping, changes in land use, jungle clearance, waste tips, etc., are endless. The ERTS also have great potential for monitoring crop health and identifying potentially hazardous situations before they develop, and so allow remedial action to be taken.

Ecohouse. ▷AUTONOMOUS HOUSE.

Ecological efficiency. The ratio between the amount of energy flow at different points along a food chain.

Ecological indicators. Organisms whose presence in a particular area indicates the occurrence of a particular set of water, soil and climatic conditions; ▷ for example, GLADIOLI and LICHENS.

Ecological niche. Each organism has a special task in an ecosystem – known as a niche. No two species of plant or animal can occupy the same niche for long; competition ensues and one species eventually adapts to occupy a different niche, or dies out.

Ecological pyramids. Diagrams which show the overall flow of energy through an ecosystem. The producer (green plant) level forms the base and the successive trophic or feeding levels (herbivores, carnivores, etc.) occupy the remaining tiers. There are various types of pyramids:

Ecological pyramids

1. *The pyramid of numbers*, which shows the number of organisms at each level of a food chain.

2. *The pyramid of biomass*, which is constructed using the total weight of organisms at each level of the pyramid of numbers. It shows the biomass at a particular time, not over a period of time. The pyramid sometimes looks as if it is the wrong way up (see Figure 13), as in the

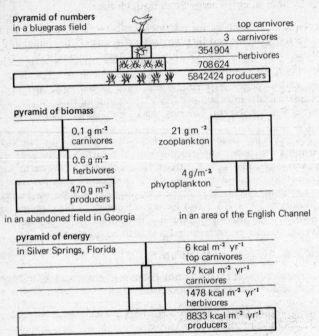

Figure 13. Ecological pyramids. A pyramid of numbers shows the totals of individual organisms found in an area at a particular time. A pyramid of biomass shows the total weight, in grammes per square metre, of organisms in an area at a particular time. A pyramid of energy shows the amounts of energy, in kilocalories per square metre per year, available to other organisms in a year's time. (From L. Pringle, *Ecology: Science of Survival*, Macmillan Co., 1971.)

case of phytoplankton and zooplankton in the English Channel. This is because the phytoplankton are eaten almost as soon as they are formed. (The rate of production of the phytoplankton is very much in excess of the growth rate of the zooplankton. The mass of zooplankton can therefore be in excess of the mass of phytoplankton, which are consumed almost immediately.)

3. *The pyramid of energy*, which presents the best overall picture of energy flow. It shows the rate at which food is produced, as well as the total amount. The pyramid is not affected by the size of the organisms, nor by how quickly energy flows through them.

Ecology. The study of the relationships between living organisms and between organisms and their environment, especially animal and plant communities, their energy flows and their interactions with their surroundings.

Economic growth. The increase in economic activity in a country, usually measured in terms of the GROSS NATIONAL PRODUCT.

It is the twin prospects of continued undifferentiated economic growth and unrestricted population growth which cause the greatest concern to those who fear for the future of our environment.

Why is it that many eminent economists continue to insist, in spite of all the growing evidence to the contrary, that there is absolutely nothing wrong with the doctrine of economic growth? Part of the answer to this question is concerned with the nature of the economic discipline itself (⟡ECONOMICS), but the following is an attempt to explain the devotion that is attached to this particular doctrine.

The recurrent nightmare of the economist, and indeed any sensible person, is a slump. In a slump men and machines are idle; no purchasing power is available to stimulate production and firms do not have the economic incentive to invest in new equipment. John Maynard Keynes asserted that to avoid a slump it is necessary to ensure that personal savings are matched by the amount that firms choose to invest (if people reduce their spending and save more, then this inevitably leads to reduced company investment in plant and eventually to a slump). His solution was to ensure that people had money to spend on the consumption of goods by means of tax cuts or public works, thus stimulating production and investment, even at the expense of running a budget deficit (⟡KEYNSIAN ECONOMICS).

These proposals became accepted by economists, for the simple reason that they appeared to work. The stimulation of consumption, if necessary by quite artificial means (advertising, planned obsolescence) is now regarded as a cornerstone of economic policy and an essential prerequisite of the slump-free economy. In effect, production and consumption have become ritualized.

However, the growth doctrine has been increasingly questioned because of its failure to differentiate between the production of useful and trivial outputs. Keynes himself was careful to distinguish between absolute needs – food, shelter, clothing, etc. – and relative or status-giving needs, the latter being potentially insatiable. It has been assumed that consumer sovereignty will prevail in maintaining a reasonable output of products that are genuinely useful. But, as Galbraith has pointed out, consumer demand is managed by producers through advertising and planned obsolescence. Consumer sovereignty turns out all too often to be producer sovereignty.

The two most serious charges against the doctrine of economic growth are (a) that it is responsible for the ritualization of the production/consumption process, which constitutes a heavy and ever-increasing drain on natural resources; and (b) that, because this ritual occurs almost exclusively in the area of insatiable (status-giving) needs, it is in this area that investment is concentrated, together with its spin-off of high profits and salaries. People outside the production/consumption ritual – nurses, teachers, many small-businessmen – find themselves falling behind workers in the ritual industries, who can always justify increased wages by claiming corresponding increases in productivity. People outside the production/consumption ritual must either see their standard of living decline or demand more for their services.

J. K. Galbraith, *The New Industrial State*, André Deutsch, 1972; Penguin Books, 1968.
J. K. Galbraith, *The Affluent Society*, Hamish Hamilton; Penguin Books, 1970.
M. J. L. Hussey, 'Has the twentieth century the technology it deserves?' *Institute of the Royal Society of Arts*, vol. 120, no. 5185, December 1971.
E. J. Mishan, *The Costs of Economic Growth*, Staples Press, 1967; Penguin Books, 1969.
For a traditional point of view, see W. Beckerman, *In Defence of Economic Growth*, Jonathan Cape, 1974.

Economics. The study of the ways in which people choose what to make of whatever resources they can get, and the ways in which they conduct and organize matters of exchange. The basic propositions of economics relate to matters of human psychology – such as 'values', offers to buy (DEMAND), offers to sell (SUPPLY), tendencies to consume or to delay consumption, and the effects of different inducements in changing these kinds of behaviour. Among those who have criticized the role that economics has come to play in our society are a number of eminent economists, including Galbraith, Mishan, Boulding and Schumacher. One can do no better than quote Schumacher in this context:

It is hardly an exaggeration to say that, with increasing affluence, economics has moved into the very centre of public concern, and economic performance, economic growth, economic expansion, and so forth have become the abiding interest, if not the obsession, of all modern societies. In the current vocabulary of condemnation there are few words as final and conclusive as the word 'uneconomic'. If an activity has been branded as uneconomic, its right to existence is not merely questioned but energetically denied. . . .

In fact, the prevailing creed, held with equal fervour by all political parties, is that the common good will necessarily be maximized if everybody, every industry and trade, whether nationalised or not, strives to earn an acceptable 'return' on the capital employed. Not even Adam Smith had a more implicit faith in the 'hidden hand' to ensure that 'what is good for General Motors is good for the United States'.

However that may be, about the *fragmentary* nature of the judgements of economics there can be no doubt whatever. Even within the narrow compass of the economic calculus, these judgements are necessarily and *methodically* narrow. For one thing, they give vastly more weight to the short than to the long term, because in the long term, as Keynes put it with cheerful brutality, we are all dead. And then, second, they are based on a definition of cost which excludes all 'free goods', that is to say, the entire God-given environment, except for those parts of it that have been privately appropriated. This means that an activity can be economic although it plays hell with the environment, and that a competing activity, if at some cost it protects and conserves the environment, will be uneconomic.

. . . it is inherent in the methodology of economics *to ignore man's dependence on the natural world.* (Excerpt from E. F. Schumacher, *Small is Beautiful*, Blond & Briggs, 1973, ch. 3.)

P. A. Samuelson, *Economics*, 3rd edn, McGraw-Hill, 1955.
J. K. Galbraith, *The New Industrial State*, André Deutsch, 1972; Penguin Books, 1968.
T. Congden and D. McWilliams, *Basic Economics: A Dictionary of Terms, Concepts and Ideas*, Arrow Books, 1976.

Ecosystem. The plants, animals and microbes that live in a defined zone (it can range from a desert to an ocean) and the physical environment in which they live comprise *together* an ecosystem. The ecosystem embraces the FOOD CHAIN through which energy flows together with the biological CYCLES necessary for the recycling of essential nutrients. Thus an ecosystem has the means of producing both energy and materials for life going on continuously. Taken on a global basis, all the separate ecosystems are the life-sustaining processes on which our survival depends. Their integrity must be preserved, otherwise biological communities can die out, or essential services not be performed, such as the self-purification of a river or the control of greenfly by ladybirds. Most ecosystems are extremely complex: for example, a deciduous forest can support over 100 species of birds, as well as many wild flowers, grasses and shrubs. Man's intervention by means of planting conifers reduces bird species, animal variety and the types of flowers, ferns and grasses. In general ecosystems are extremely resilient and where a consumer (◊ FOOD CHAIN) has more than one source of food, then if one becomes less readily available the consumer can still survive. However, if the consumer depends on or only has one source of food available, then its survival can be in jeopardy.

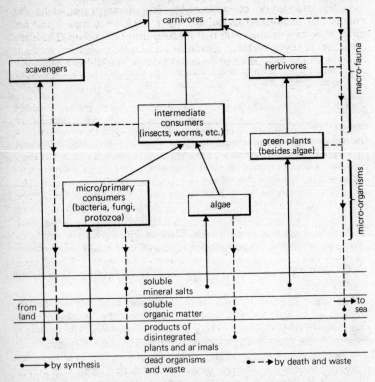

Figure 14. A simplified aquatic ecosystem for a river-bed community.

Figure 14 shows an aquatic ecosystem where the primary inputs are energy, by photosynthesis fixed by the primary producers algae and green plants, *plus* the organic and inorganic materials carried by the river. The primary and micro-consumers (◊ DECOMPOSERS; PROTOZOA) eat dead organic matter; the intermediate consumers, worms and insect larvae, feed on the primary consumers and algae. The herbivores eat the green plants. The carnivores (fish) eat the intermediate consumers. The scavengers eat the bottom debris and dead organic matter too, so that virtually nothing is wasted. Thus, the ecosystem is a complex interlinking arrangement with its own form of *equilibrium*, i.e. a balance is struck between the total production of living material and the rate of death and decay over a period of time. This is ecological equilibrium. The more complex the ecosystem, the greater the stability.

Man's intervention often simplifies ecosystems dramatically, for example, by mono-crop agriculture where some fields grow wheat for up to 30 years with the result that diseases tend to build up. These are often kept at bay by synthetic chemicals whereas the traditional rotation of crops would have ensured (usually) that disease did not build up. Other common examples of the dangers involved in man's modification of ecosystems are the conversion of prairie land to wheat, which destroys grassland ecosystems, the logging of deciduous forests to leave barren land which can lead to soil erosion, or, if replanted with conifers, to a modification of forest ecosystems. Large-scale changes in the use of land also result in a change in the ALBEDO, often increasing it, which in turn affects the MEAN ANNUAL TEMPERATURE and thus cumulatively the earth's climate. The spraying of crops with insecticides can be a major violation of an already simplified ecosystem. The animals at the top of the food chain are smaller in number than those below. Predators which eat pests, e.g. carnivores, are more likely to be destroyed in a blanket-spraying programme compared with the pests as there are many more pests than predators. Thus, once spraying is initiated and the equilibrium upset, it can become a self-perpetuating cycle, spraying then becoming a necessity as the predators have been destroyed. Pests seem to develop resistance to pesticides more quickly than their predators (◇DDT).

In accepting the undoubted short-term benefits of modifying ecosystems, we may in the long term lay ourselves open to the risk that the original ecosystems have been destroyed or gene banks of *naturally* resistant crops or animals lost for ever. (◇GENETIC EROSION.)

Effective chimney or stack height. The effective height of a chimney is the sum of its actual physical height and the rise of the emitted plume caused by its buoyancy and EFFLUX VELOCITY. The effective chimney height is used in air pollution dispersion calculations.

Effluent, Physico-chemical treatment. The effluents from food-processing industries are often extremely noxious and costly to treat; yet they contain fats and proteins (soluble and insoluble) whose recovery can yield valuable products. The separation of the fats and proteins can be done by flotation and flocculation followed by ION EXCHANGE, i.e. by physico-chemical means. One such plant is shown in Figure 15 for the treatment of poultry-processing wastes.

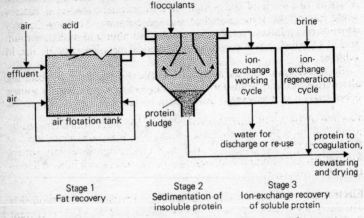

Figure 15. Physico-chemical treatment of livestock processing wastes for fat and protein recovery.

The incoming effluent is acidified to pH 3 (◊ pH) and a coagulating agent added, so that when air is bubbled through the first chamber the fat floats and is skimmed off. The protein-rich liquor is neutralized (pH 7) and flocculants added which make the insoluble protein 'floc' or come together. The protein then settles and is sent for stabilization. The soluble protein can only be removed by ion-exchange techniques and the problem is to obtain the correct material for the solution to be treated.

Cellulose ion-exchange media can selectively remove the high molecular weight protein molecules and the final clear water from the plant can be re-used or discharged. The ion-exchange medium must be recharged by brine which removes the protein molecules from the medium in a protein-brine solution which can then be coagulated and dried. The recovered protein has potential for animal feedstuffs, thereby recovering a valuable by-product and reducing effluent biochemical oxygen demand by as much as 90 per cent.

Paul Butler, 'Processes in action', *Process Engineering*, May 1975, p. 6.

Efflux velocity. The speed at which gases are emitted from the top of a chimney. It is a major design parameter in air pollution dispersion.

Electric car. One of the alternatives to the internal combustion engine for private-car propulsion. At present, electric cars have a range of about 60 miles before recharging because of the characteristics of lead-acid batteries. A lightweight sodium-sulphur battery is being developed which is one-tenth of the weight of an equivalent lead-acid battery while having the same electrical storage capacity. The new battery depends on an interaction between molten sulphur and molten sodium at 350°C, which poses considerable practical problems for its use in vehicles.

Electric cars are seen as potentially viable short-range non-polluting alternative means of transport by the mid-1980s but the energy source (electricity) is still generated in the main by fossil fuels.

Electrodialysis. The separation of ionic components from solutions by means of ion-selective membranes under the influence of an electric field. Electrodialysis is used in DESALINATION of low-salinity waters for human consumption, kidney machines, and specialized effluent-treatment processes.

Electrolysis. The chemical splitting of an electrolyte – a solution that can carry an electric current – by the passage of an electric current through it. The solution is ionized into positively and negatively charged IONS which move towards the oppositely charged electrodes immersed in the solution. Once at the electrode they give up their charge and can be collected. Thus, the electrolysis of water gives HYDROGEN and oxygen as separate components. The hydrogen can then be used in a FUEL CELL.

The electrolysis of brine – sodium chloride (NaCl) and water (H_2O) – gives free CHLORINE (Cl) plus hydrogen (H_2) plus a solution of caustic soda (NaOH). The electrolysis of brine (◇CHLOR-ALKALI PROCESS) is the main process for chlorine manufacture.

Electromagnetic conservation. An aspect of the general problem of conservation of scarce resources little known to most people is the need to conserve the electromagnetic spectrum. All radio and television stations occupy a certain part of the electromagnetic spectrum – how much depends on technical factors – and yet the total available spectrum is strictly limited. All aspects of economic growth for the last half-century have been very closely linked to growth in the use of radio communication, not merely broadcasting, but also, and in some ways more importantly, point-to-point communication. If the electromagnetic spectrum became completely congested, this would have a major impact on world economic development.

The available electromagnetic spectrum is limited by inescapable physical factors which can never conceivably be overcome other than by the invention of some means of communication which simply does not depend on electromagnetic waves. That, at the moment, seems an extremely remote possibility.

One way of moderating the increasing problems in the scarcity of raw materials and energy is by the effective use of radio communication. For example, it can be demonstrated that on average the introduction of mobile radio into a commercial vehicle can reduce its fuel consumption by about 20 per cent. On a more ambitious scale, one can imagine the savings in energy and transport costs that could be obtained by giving office staff communication links (video-phone, telex, computer terminals) to permit them to work effectively at home instead of travelling to and from work every day.

The situation is not a hopeless one since it is clear that technical resources currently known to us can greatly improve our use of the spectrum. However, if steps are not taken fairly urgently, then the ability of radio to contribute to the solution of the technological problems of the late twentieth century will be inhibited.

Electron. A negatively charged elementary particle which is present in the orbital structure of all ATOMS.

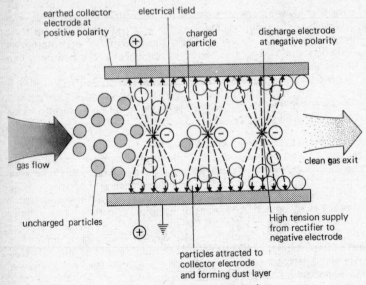

Figure 16. The principles of electrostatic precipitation.

Electrostatic precipitator. The gases from combustion processes contain large burdens of particulate matter, grit, smoke, dust and fume. Before discharge to atmosphere they must be cleaned. An electrostatic precipitator is one method. The dirty gases are passed through an intense electric field and become electrically charged as shown in Figure 16. The charged particles are attracted to the collector electrodes which have an opposite polarity where they accumulate. A mechanical handling or rapping system dislodges the accumulated dust which by this time has agglomerated and it falls out of the gas stream. The product from power stations is called fly ash and is used in a variety of land reclamation and building materials manufacture.

Electrostatic precipitators can remove 97 to 99·5 per cent of the initial grit burden. They can also be used to remove acid or oil mists and fumes and are thus extremely versatile and efficient gas-cleaning devices.

The effect of collection efficiency is shown in Figure 17 and shows the

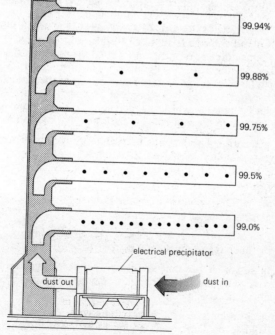

Figure 17. Dust collection efficiency.

91

importance of fractional increases in efficiency in dust-collection equipment, e.g. there are roughly 250 000 tonnes of fly ash emitted from power stations in the UK at an approximate precipitator efficiency of 97·5 per cent; if this could be raised to 98·75 per cent the emissions would be *halved*. Precipitators of 99·7 per cent efficiency are now used on new installations.

Element. A simple substance which cannot be broken down into simpler substances by chemical methods. An element consists of ATOMS all with the same ATOMIC NUMBER.

An element may be represented by a chemical symbol which is used in formulae and equations. The symbol is composed of the initial letter of the name, plus another letter where necessary. Usually the initial letter of the English name is used, but sometimes it is that of the Latin name. The symbol is used to represent an atom of the element as well as the element itself. An abbreviated table of elements is included at the front of this book.

The symbols representing the elements may be combined to form chemical formulae which represent compounds: for example, sodium chloride (formula $NaCl$) consists of equal numbers of atoms of sodium (symbol Na) and chlorine (symbol Cl). When a molecule of the compound contains different elements in unequal proportions, their ratios are indicated by subscripts: for example, water (formula H_2O) consists of hydrogen (symbol H) and oxygen (symbol O) in the ratio 2:1, two atoms of hydrogen joined to one atom of oxygen. These are examples of molecular formulae in which the individual and total number of the constituent atoms in the molecule are indicated. (◇ISOTOPE.)

Emission standard. The amount of pollutants permitted to be discharged from a pollution source. Emission standards are commonly described in one or more of the following ways for liquid, solid or gaseous pollutants:

1. Mass of pollutants over a certain time: e.g. pounds per hour, kilogrammes per hour or tonnes per day.

2. Mass of pollutants per unit mass of material processed: e.g. 40 pounds per tonne (approximately 2 per cent).

3. Mass of pollutant per unit volume of discharged gas at specified conditions of temperature and pressure: e.g. grains per cubic foot or milligrammes per cubic metre.

4. The concentration specified as the volume of pollutants (if gaseous) per unit volume of discharged gas, again at specified conditions of temperature and pressure: e.g. parts per million.

The emission standards for NOISE and IONIZING RADIATION are dealt with under their respective entries. (◊ THREE-MINUTE MEAN CONCENTRATION; THRESHOLD LIMITING VALUE; CONCENTRATION.)

Endosulfan. An ORGANOCHLORINE insecticide. It is toxic to fish in concentrations as low as 0·00002 parts per million, i.e. 1 million (UK) gallons would require 0·0032 oz of Endosulfan, if evenly distributed, to render it toxic to fish. A classic case illustrating the toxicity of organochlorines occurred when a few gallons of Endosulfan spilled into the Rhine in June 1969, causing the death of many million fish.

Endrin. ◊ CHLORINATED HYDROCARBONS; ALDRIN.

Energy. The ability to do work. Energy has many forms: mechanical, chemical, thermal (heat), nuclear, electrical.

It is common practice to consider energy from thermal sources, such as the heat released from the combustion of fossil fuels or nuclear fission, to be primary energy as it can only be used for heating and must be converted by means of an engine such as a steam turbine before it can do work. The conversion of heat to electrical or mechanical energy is governed by the second law of thermodynamics (◊ LAWS OF THERMODYNAMICS).

Approximate energy conversion ratios for heat to work for a thermal power station range from 20 per cent (poor) to 40 per cent (excellent). Thus a 33 per cent efficient power station requires an input of 3 kilowatt hours thermal (3 kW h_{th}) for 1 kilowatt hour electrical (1 kW h_e) output. Thus the form in which energy is supplied must be specified when ENERGY ANALYSIS is being performed. A process which uses 2 kW h_{th} per unit output is much more energy efficient than a competitive process which requires 1 kW h_e per unit output, as the competitor process in effect requires a thermal input at the power station of about 3·3 kW h_{th} to make 1 kW h_e. (◊ ENERGY RESOURCES; WATT; JOULE.)

Energy analysis. Every operation carried out on materials, in the mining of fuels or metals, refining or working of metals, transporting, manufacturing and construction of all kinds, involves the use of energy. Hence, any object, from a power station to a milk-bottle, from a barrel of oil to a newspaper or a hole dug in the ground, requires the expenditure of energy to bring it into existence – and also to dispose of it.

It is possible to examine the details of the processes by which something is made and to determine the amount of energy needed to produce it. This involves adding up the amounts of energy needed to make each part and that needed to assemble the product. For each part the energy required to form it and the energy required to provide the necessary materials are included, and allowances must be made for some fraction of the energy needed to make any machines involved in the various processes. This is called 'energy analysis by process analysis'.

Other methods, usually less accurate since they involve the use of non-physical variables such as financial costs, may be used for energy analysis.

Energy analysis constitutes a valuable supplement to economic analysis when attempts are made to estimate the effects of changes in the availability of energy upon a national economy. Energy analysis is based extensively upon physical reasoning; this enables it to evade many of the difficulties that confront economic analysis, and which derive from such matters as absolute and relative variations in price. Nevertheless, energy analysis can only be carried out in accordance with one or another of several possible sets of conventions – concerning, for example, how much of the energy expended should be attributed to each of the products of a process with several outputs. It is vital, if intelligible comparisons are to be made and confusion avoided, that great care is taken to state the conventions under which an energy analysis is made.

S. Nilsson, 'Energy analysis', *Ambio*, vol. 3, no. 6, 1974.

Energy conservation. The rational use of ENERGY RESOURCES is now a pressing problem and embraces a wide range of options ranging from RECYCLING of non-fuel materials to the avoidance of competition between various sources such as coal, gas and oil, and the adoption of a rational fuel policy. The areas where improvements can be made or new techniques added are shown in Figure 18.

One domestic improvement is cavity-wall and roof insulation which, if implemented to the full, would cut domestic energy consumption by 25 to 30 per cent. The scope for improvement is large indeed but there has been no real attempt in government or industry to conserve energy to any significant degree.

The reason for this apathy may lie in the lack of financial (or other) incentive for business and individuals to save energy. Air-conditioned buildings in London have an annual rental of £10 per square foot per annum, shortly to increase to around £20 per square foot per annum and who knows where after that. The running costs of a moderately designed air-conditioning installation are 40p to 60p per square foot per annum and the depreciation costs are of the same order. Thus, the financial incentive to reduce both capital and running costs and conserve energy is virtually non-existent as the marginal savings would amount to approximately 5 to 10 per cent of the rental; hence inefficient energy-intensive air-conditioning can and is used. As long as energy is

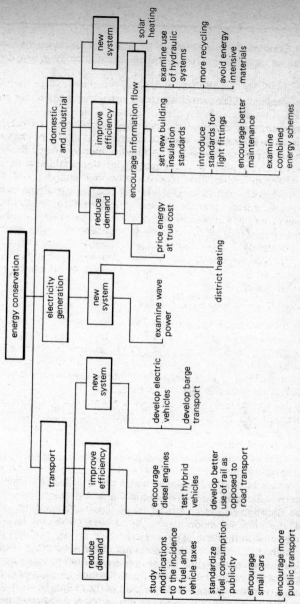

Figure 18. Proposals for energy conservation. (From *Energy Conservation*, HMSO, 1974.)

a cheap commodity in relation to the value of the goods or services provided or manufactured by it, then this attitude may persist.

Energy cropping. ⟡CELLULOSE ECONOMY.

Energy demand. The way western society consumes its energy resources is illustrated in Figures 19 and 20, which show the different

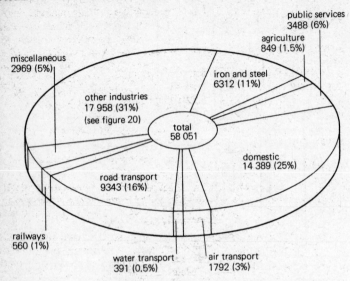

Figure 19. Energy consumption by final users in 1972 on the basis of heat supplied (million therms). (From *Energy Conservation*, HMSO, 1974.)

levels of energy consumption by energy and by final users in the United Kingdom in 1972.

It is precisely those technologies which are regarded as giving us our productive efficiency, measured in economic terms, that are responsible for the alarming rate of growth of energy consumption in the western world. For example, modern agricultural methods require large fossil fuel subsidies (in the form of OIL, PESTICIDES, artificial FERTILIZERS, etc.). In the USA this has been estimated (Perelman, 1972) as equivalent to an input–output ratio of 5:1 in energy terms (⟡AGRICULTURE, ENERGY AND EFFICIENCY ASPECTS). Furthermore, as populations continue to grow, demands for increased food output will require large increases in the use of nitrate fertilizers (⟡GREEN REVOLUTION).

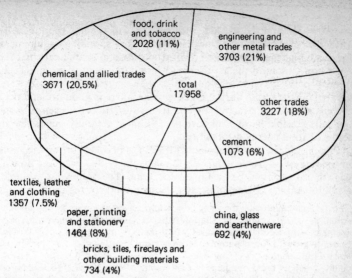

Figure 20. Energy consumption of 'other industries' (see Figure 19) of the basis of heat supplied (million therms). (From *Energy Conservation*, H M S O, 1974.)

But present methods require about 2 kilogrammes of coal equivalent to fix 1 kilogramme of nitrogen. The energy demands of sea water DE-SALINATION plants, required for increasing water production in the many arid areas of the earth, are also very high.

Future prospects are of increasing energy demands as essential industrial materials become scarce. The following table gives the total direct energy inputs for producing certain vital metals, neglecting transport costs for both products and raw materials. Bear in mind that the combustion of coal yields about 29 megajoules per kilogramme (MJ/kg):

	Megajoules per kilogramme
Copper from 1·0 % sulphide ore in place (1940s)	~ 54
Copper from 0·3 % sulphide ore in place (1980s)	~ 98
Aluminium from 50 % bauxite in place (1970s)	~ 204
Magnesium from seawater (anytime)	~ 360
Titanium from ilmenite in place (1970s)	~ 593

From J. C. Bravard, H. B. Flora II, and C. Portal, *Energy Expenditures Associated with the Production and Recycle of Metals*, O R N L–N S F–E P–24, Oak Ridge National Laboratory, November 1972.

As these ores become progressively more scarce ($\diamond$MATERIALS RESOURCES) and the market price rises, it will become economically attractive to work leaner sources, with a corresponding requirement for much higher energy inputs. Unfortunately, it appears likely that for most non-structural metals there is not a continuum of progressively poorer ores in growing quantities ($\diamond$ARITHMETIC–GEOMETRIC RATIO) but rather an abrupt grade gap, i.e. there is good ore and then there is rubbish. Extracting metals from this 'rubbish', even if economically feasible because of scarcity, is almost certainly impossible in terms of total energy availability.

Present trends, therefore, would all seem to point to further increases in the rate of future demand. Yet it seems likely that even our present growth rate cannot be sustained for long. The following example, given by Amory Lovins, illustrates the magnitude of the problem:

The aggregate amounts of energy now being converted are so prodigious that voluntary rapid change in supply patterns is *physically impossible*. For example, suppose that our present world conversion rate of 8×10^{12} W continues to grow (as most authorities predict and urge) by about 5%/yr for the rest of this century, yielding a $3\cdot7\times$ increase to about 3×10^{13} W. If we could somehow build one huge (1 GW $= 1000$ MW$_e = 10^9$ W$_e$) nuclear power station per *day* for the rest of this century, starting today, then when we had finished, *more than half* of our primary energy would still come from fossil fuels, which would be consumed about *twice* as fast as now. This is an optimistic case, in that it is hard to think of any knowledgeable person who thinks such a rapid nuclear infusion is possible, even were it advisable.

A. B. Lovins, *World Energy Strategies*, Earth Resources Research for Friends of the Earth, 1973.
M. J. Perelman, 'Farming with petroleum', *Environment*, vol. 14, no. 8, October 1972, p. 8.

Energy, Effect of conversion on climate. The influence of man-made heat as a result of our energy conversion activities is already significant on a local scale and will soon become significant regionally if present trends continue. Most large industrial areas already add 10 per cent to the sun's heat input, producing significant local variations in climate, and such areas are likely to grow and proliferate.

An extreme example of such 'heat islands' is Manhattan Island in New York City, with a man-made power density of over 700 watts per square metre compared with 93 watts per square metre from the sun ($\diamond$TEMPERATURE INVERSION).

On a global scale, significant changes in climate may be triggered by relatively small variations in the heat balance in critical areas such as the floating Arctic pack ice. At present growth rates, one can foresee the possibility of such man-made changes in the next century, resulting from the combined effects of increases in man-made heat, man-made particulates in the atmosphere (Figure 21), and increased levels of carbon dioxide (Figure 22).

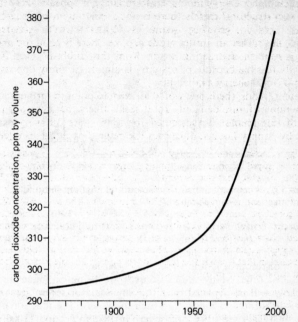

Figure 21. Changes in carbon dioxide concentation. (From Machta, 1971.)

It is unfortunate that most of the basic questions about climatic mechanisms are at present unresolved, and it is therefore almost impossible to make assertions on sound scientific bases regarding the specific climatic effects of human activity. However, present rapid growth rates are viewed with considerable disquiet by many climatologists, and a prudent policy of energy conversion on a global scale, while undoubtedly impossible politically, is evidently urgently required.

See A. B. Lovins, *World Energy Strategies*, Earth Resources Research for Friends of the Earth, 1973.

Energy resources. There are two forms of resources, income and capital or renewable and non-renewable. Once a capital resource is spent, it cannot be recovered; oil, coal and uranium are capital

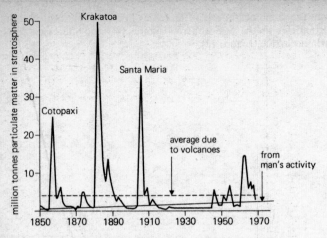

Figure 22. Time variations in the amount of dust put into the the atmosphere. (From Mitchell, 1970.)

resources. Solar energy is an income source and comes in continually. The derivatives of solar energy such as grass, trees, etc., are forms of income but are finite in the rate at which they can be exploited. A finite or capital resource is one which has been laid down or formed over geological time. Many such resources are being exploited and quite probably will be depleted over the time-scale of several generations. This is illustrated in Figure 23 which shows world production of crude oil, which has risen from almost zero in 1880 to around 23 000 million barrels per year in 1973 and is an example of exponential growth (◊EXPONENTIAL CURVE). When the production rate is divided into the amount of the proved available reserves in the ground, the RESERVE–PRODUCTION RATIO is obtained. Figure 24 shows the curve of the upper and lower estimates for world crude-oil production based on the exponential increase in Figure 23 for values of ultimate total oil production of 1350×10^9 and 2100×10^9 barrels respectively. Eventually the rate of production cannot be increased any more and rate of discovery is less than production and so the production will decline to zero – in other words, it is a finite cycle. Figure 24 shows that the middle 80 per cent of oil production on the lower estimate (1350×10^9) would occupy a mere 58 years; for the more optimistic estimate (2100×10^9) it would occupy 64 years.

The long-term availability of usable sources of energy and our ability to use them wisely are the factors which will be predominant in

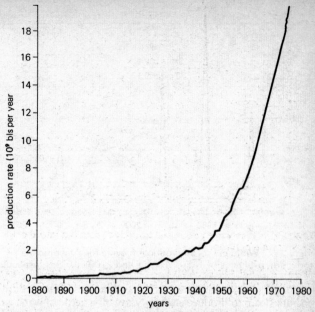

Figure 23.   Global production of crude oil since 1880.

deciding the length and nature of the human race's habitation of our planet. In considering the availability of energy sources one must remember that social, economic, political and geographical factors can be as important as the actual abundance or scarcity of a particular resource. As an example, it seems likely that Britain's North Sea oil reserves, instead of being properly husbanded, will, because of economic demands (a guaranteed minimum return on capital invested) be used up as rapidly as they can be extracted (alternatively they may be sold or mortgaged before we have even got them). Such policies appear quite rational to the conventional wisdom in economic policy, which tends to regard money and goods as equivalent. Economic attitudes to energy sources also disguise the very nature of these resources, i.e. that they are a *degradable* resource which, to be used effectively, must take into account the continuous chains of conversion from a high-energy form to low-temperature heat (e.g. car exhausts). It is economic considerations which prevent, for example, the Central Electricity Generating Board from designing power stations to make the most thermodynamically efficient use of their waste heat (◊DISTRICT

101

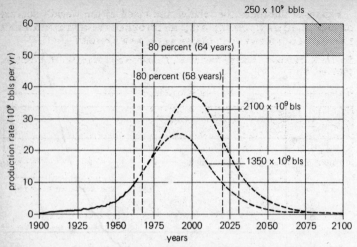

Figure 24. World crude-oil production cycle for total ultimate production estimates of 1350×10^9 bls and 2100×10^9 bls respectively.

HEATING). And in energy terms it hardly makes sense to burn gas to generate steam to produce electricity to boil a kettle of water or heat a house, when those operations could be carried out so much more energy-efficiently by using gas directly.

At present, mankind is using energy on a global scale at a rate approximately equivalent to 20 times all the energy associated with the food-gathering processes required by the human race (agriculture, fishing, hunting and gathering). Furthermore, our rate of increase of energy conversion is almost three times that of population growth (i.e. about 5·7 per cent per annum). However, the way in which we use our energy supply is extremely patchy. The USA, with only 6 per cent of the world's population, uses over one-third of the world's energy. This gross inequality of distribution is one of the most disturbing features of the present situation. It has been said (by R. O. Anderson, Chairman of Atlantic-Richfield) that 200-odd million Americans use more energy for air-conditioning than 800-odd million Chinese use for everything! Further, it is in those countries with the largest GROSS NATIONAL PRODUCT that the rate of growth of energy consumption is greatest due to increased material standards of living and the use of energy-intensive technologies. It is these countries whose economies are at greatest risk when the realities of energy limitations become apparent.

Figure 25 illustrates the relative use made of the world's sources of energy. (⟡GAS; COAL; OIL; HYDROELECTRICITY; NUCLEAR ENERGY; SOLAR ENERGY; OIL SHALES; TAR SANDS.)

See M. K. Hubbert, 'Energy resources', *Resources and Man*, W. H. Freeman, 1969, ch. 8.

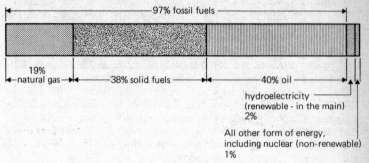

Figure 25. Relative uses made of energy sources. Note that this diagram represents the use of industrial energy and that, in large areas of the world, people and animals are the predominant source of energy.

Environment. All of the surroundings of an organism, including other living things, climate and soil, etc. In other words, the conditions for development of growth.

Environmental resistance. The sum of the factors which keep populations from reaching their maximum potential, e.g. disease, predators, competition, climate, etc. The human population has increased greatly because ways of reducing the environmental resistance have been found. Most populations in nature do not show high rises or great falls. There is usually a balance between biotic potential and environmental resistance.

Enzymes. Proteinaceous substances that catalyse microbiological reactions such as decay or fermentation (⟡CATALYSIS). They are not used up in the process but speed it up immeasurably. They can promote a wide range of reactions, but a particular enzyme can usually only promote a reaction on a specific substrate. (⟡ENZYMES IN THE HUMAN ORGANISM; ENZYME TECHNOLOGY.)

Enzymes in the human organism. The metabolic processes of living cells are brought about by interlinked systems of enzymes produced in various tissues of the body. In general, enzymes can be regarded as

103

acting in series, with the product of the action of one enzyme forming the substrate of the next.

The digestive and respiratory processes, glucose metabolism, ammonia removal and muscle energy production are all catalysed by enzymes. Digestion is effected by amylolytic, lypolytic and proteolytic enzymes breaking down starches, fats and proteins. The various enzymes are secreted by organs of the digestive system such as the stomach, intestines, liver and pancreas. Respiratory enzymes present in tissues of the body take part in the oxidation-reduction system involving molecular oxygen. Similarly by the action of enzymes, the urea cycle provides for the removal of waste products, especially ammonia produced by the metabolism of amino acids.

Two very important metabolic systems, the Krebs cycle and the Embden–Meyerhof pathway, are catalysed by numerous different enzymes. The Krebs cycle is the major pathway for the breakdown of carbon chains in sugars, fatty acids and amino acids, and also is the main source of energy for other enzyme pathways. The E–M pathway is responsible for glucose metabolism.

In addition to these examples, many other parts of the body such as the brain and the muscles secrete enzymes which aid the various functions and processes necessary for life in the human organism.

Enzyme technology. Enzyme technology covers a wide commercial area which embraces drug manufacture, beer malting, stabilization of foods, and protein production. Thus the enzymes known as cellulases ('-ase' indicates an enzyme; '-ose' is the substrate or product, e.g. cellulose) are the class of enzymes that promote the decomposition of cellulose, in this case to glucose. There are some 13 000 micro-organisms that live on cellulose and they accomplish this by secreting enzymes which break down the substrate for a food source. One organism (*Trichoderma viride*) can produce a cellulase rich in a component capable of breaking down crystalline cellulose.

Using cellulose as an example, a commercial process for cellulose exploitation (♢CELLULOSE ECONOMY) requires that the micro-organism (in this case *Trichoderma viride*) which produces a large amount of cellulase is grown on a culture medium containing spruce pulp and nutrient salts. The culture is filtered leaving an enzyme solution which is then available to hydrolyse cellulose substrates, that is, convert the cellulose to sugars (♢HYDROLYSIS). So far tests have been conducted on milled newspapers; the cellulase solution and newspapers are placed in a reaction vessel under strictly controlled conditions (pH 4·8, 50°C). The time taken to break down the cellulose is of the order 20 to 80 hours and a product of glucose syrup obtained. Any unutilized cellulose is recycled along with the enzymes. The yield of glucose is roughly 50 per cent of the original cellulose. The glucose is then used for FERMENTATION products and/or the production of SINGLE CELL PROTEIN. Figure 26 shows a flow diagram for the process based on the US Army Natick Laboratory work.

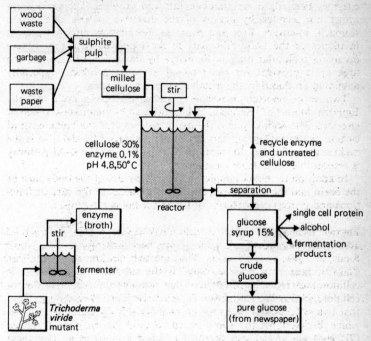

Figure 26. Natick process for enzymatic hydrolysis of cellulose to glucose.

Enzymes require strictly controlled conditions and are easily destroyed or inactivated by many substances. Feedstock purity and processing conditions must be closely monitored. Although not consumed in the reaction, they eventually degrade and must therefore be continually synthesized. In the cellulase reaction, product inhibition can eventually occur and the reaction cease, with the result that all the cellulase cannot be used in the one go. Multi-charging must then be used if all the cellulose is to be converted to glucose.

This area of enzyme technology has great promise for the production of single-cell protein from low-grade starchy materials, thereby upgrading animal feedstocks.

Epidemiology. The study of categories of persons and the patterns of diseases from which they suffer so as to determine the events or circumstances causing these diseases. If a cause is discovered, then those responsible for PUBLIC HEALTH policy can take appropriate steps to prevent the disease in question.

A classic case of epidemiology was Snow's identification of contaminated water from the Broad Street pump in 1854 as being responsible for the spread of CHOLERA.

Another example of the role of epidemiology is the evidence that smoke and sulphur dioxide act in both causing and aggravating bronchitis. The Clean Air Acts brought a striking reduction in extra deaths due to these pollutants as shown in the table below for London.

Year	Maximum daily concentration (grammes per cubic metre)		Extra deaths in Greater London
	Smoke	Sulphur dioxide	
1952	76000	3500	4000
1962	3000	3500	750
1972	200	1200	Nil

From Royal Commission on Environmental Pollution, Fourth Report: *Pollution: Progress and Problems*, Cmnd 5780, HMSO, December 1974.

The epidemiology sifts the evidence, identifies patterns and causes, and we know with reasonable certainty that the Clean Air Acts have done the job they were intended to do.

J. J. Snow, *On the Mode of Communication of Cholera*, 2nd edn, Churchill, 1885.

Erosion. The lowering of the land surface by weathering, corrosion and transportation, under the influence of gravity, wind and running water. It can also apply to the eating away of the coastline by the sea.

Estuarial storage. The storage of water for domestic water supplies in estuaries when suitable inland sites for reservoirs have been exhausted. In the UK, Morecambe Bay, the Solway Firth and the Wash have been suggested as potential sites, with a barrage used to control or prevent the entry of sea water.

Ethanol. Ethyl alcohol (C_2H_5OH). It has a boiling point of $78.4°C$ and a specific gravity of 0.789. Ethanol is mainly synthesized from ethylene in the petrochemical industry. It can also be made by the

FERMENTATION of sugars, or by HYDROLYSIS of refuse or celluose waste.

It can be used as a motor fuel, and is sold under the name 'power alcohol'.

Eutrophication. The natural ageing of a lake or land-locked body of water which results in organic material being produced in abundance due to a ready supply of nutrients accumulated over the years. Eutrophication can be greatly increased by man as a result of nitrates from fertilizer run-off (◊MODERN FARMING METHODS) and sewage treatment processes (◊SEWAGE EFFLUENT TREATMENT).

A eutrophic lake is highly productive in organic material and can result in algal blooms (◊ALGAE) which are short lived and whose decay imposes a heavy oxygen demand on the water (◊DISSOLVED OXYGEN). Thus nutrients from sewage and fertilizers can ruin a lake and cause the loss of a body of water which may be of use to man. This has happened in parts of the Great Lakes system where it has been estimated that Lake Erie received 37 500 tonnes of nitrogen from run-off and 45 000 tonnes from sewage in 1968 alone. The result of these man-made nutrients is supposed to have aged the lake 15 000 years quicker than if it were left to its own devices. (◊OGILTROPHIC.)

Exponential curve. In many natural phenomena, in which, say, yeast organisms are reproducing, their rate of increase is not constant but rather the rate of increase itself is continually increasing because the organisms themselves double in number in say 10 hours and double again from the 10-hour number in 20 hours. Thus, at time zero, if the number of organisms per cubic metre is 1 million, then in 10 hours there will be 2 million, in 20 hours 4 million, in 30 hours 8 million and in 40 hours 16 million.

Such a rate of growth is called exponential growth and a curve such as Figure 27 is known as an exponential growth curve. It is plotted as

$$y = e^{kt}$$

where y is the number measured at time t, and k is a constant of proportionality determined by experiment; e is the base number of what are known as natural LOGARITHMS and has the value 2·7183.

Using world population as our model, let

N_0 = population at time zero
N_t = population at time t
k = growth constant
t = time in years.

Then $$N_t = N_0 e^{kt},$$

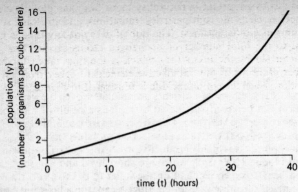

Figure 27. Exponential growth curve for the number of organisms doubling every ten hours.

i.e. population at any time t is a function of the population at time $t = 0$ (that is, N_0) and the exponential constant k.

If we wish to know how long it will take for the world population to double given that the annual growth rate is 2 per cent (this gives us k), we proceed as follows:

$$N_t = N_0 e^{0.02t}$$

and for population doubling

$$\frac{N_t}{N_0} = 2,$$

that is,

$$2 = e^{0.02t}.$$

Using natural logarithm tables where the natural logarithm of e is 1, we find out what value of (0·02t) gives us 2:

$$\text{nat. log } 2 = 0.02t$$

From the tables, nat.log $2 = 0.6931$, therefore

$$t = \frac{0.6931}{0.02} = 34.65 \text{ years.}$$

Thus the doubling time for a 2 per cent increase is 35 years.

We have dealt with exponential growth, but we can have exponential decay as well, which is represented by the equation

$$y = e^{-kt}$$

and is just as important in natural systems. For example, the atoms of radioactive elements emit particles (radiation), and in doing so decay to atoms of a different mass. The rate of decay at any time is proportional to the total number of unchanged atoms at that time and is characterized by the HALF-LIFE which is the time taken for half the number of atoms of any radioactive element to decay from an initial condition. The concept of half-life is illustrated in Figure 28.

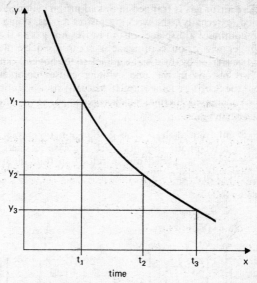

Figure 28. Exponential decay of a parameter y. The figure shows a graph of $y - e^{-kt}$ upon which three points, (t_1, y_1), (t_2, y_2) and (t_3, y_3), are picked out. The values of y_1, y_2 and y_3 are related by $y_1 = 2y_2$ and $y_2 = 2y_3$. The properties of the exponential curve are such that for this case $t_3 - t_2 = t_2 - t_1$. In other words, the time taken for y to be reduced to half its initial value is the same, whatever that initial value y is taken to be.

Exposure—dose effect relationships. The effect of exposure to a pollutant is a function of the pollutant, the type of target (e.g. animal or vegetable), and the concentration of pollutant and duration of exposure.

Short-term exposures to high concentrations are not usually equivalent to one-hundredth of the concentration for 100 times as long. The low-concentration long-period exposure may well have minimal effect, whereas the former may have a serious effect. The converse can also be true in some circumstances.

External combustion engine. An engine where the heat source or fuel combustion is outside the engine as opposed to inside the engine as in the INTERNAL COMBUSTION ENGINE. External combustion allows a wide variety of fuels to be used efficiently to supply thermal energy to the engine. However, this class is mainly dominated by steam turbines and steam engines which are hardly suitable for motor cars but very suitable for ships, power stations, etc. An exception is the Stirling engine, where air or gas is trapped in a dual piston cylinder. When the gas is heated (externally), the working piston moves, doing work. As the motion continues a displacer piston moves hot gas to the cool end of the cylinder where, on cooling, it is compressed by the working piston and transferred by the displacer back to the hot end. This method of using two pistons in the one cylinder causes complexities and expense, but the Stirling engine is quiet, virtually non-polluting and has prospects of increased thermal efficiency compared to the internal combustion engine.

Extraction of oil from shales. The gasification or extraction of oil from shales is a technique that may well increase as energy reserves decline. The process is carried out in retorts in four separate stages (from the bottom of Figure 29).

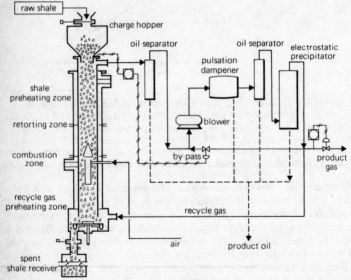

Figure 29. Gas production from oil shales.

1. The recycled gas stream is preheated by the spent shale.

2. The gas is ignited and the heat of combustion causes the oils in the shale to be driven off in the retorting zone.

3. The hot gases and oils are used to preheat the incoming shale.

4. The product gases and oils are cleaned in oil separators and an electrostatic precipitator is used to remove oil mist if necessary.

Part of the gas is recycled to run the retort and the remainder plus the oil is sold.

For questions concerning the applicability of the process on a scale sufficient to provide a significant amount of oil, ⬦OIL SHALES.

F

Fats. ◊NUTRITIONAL REQUIREMENTS, HUMAN.

Feed–conversion ratio. The ratio of weight of feed to gain in weight (in kilogrammes) in fattening cattle or poultry, etc. Typical values lie between 3 and 10, depending on animal and feed analysis. The use of hormones to promote growth is one method of reducing the feed–conversion ratio. However, one hormone, diethylstilboestrol, is a suspected carcinogen. Even if this is not the case, it raises the question of just how far man can tamper with natural systems and living creatures for short-term gains but with unknown long-term consequences.

Fermentation. The decomposition of organic substances by micro-organisms and/or enzymes. The process is usually accompanied by the evolution of heat and gas, and can be AEROBIC or ANAEROBIC. Examples:

Alcoholic fermentation (anaerobic)
$$C_6H_{12}O_6 = 2C_2H_6O + 2CO_2$$
sugars   ethyl   carbon
             alcohol   dioxide

Lactic fermentation (aerobic) – a common process for the production, preservation and seasoning of food
$$C_6H_{12}O_6 \rightarrow 2C_3H_6O_3$$
sugars   lactic acid

Acetic fermentation (aerobic)
$$C_2H_6O + O_2 = C_2H_4O_2 + H_2O$$
ethyl   acetic
alcohol   acid

Aerobic fermentation of glucose solutions (and other substrates, e.g. hydrocarbons) leads to yeast growth which can then be harvested as a source of SINGLE CELL PROTEIN.

The crucial parameters in industrial aerobic fermentation are the oxygen requirement per tonne of substrate and the removal of the heat evolved during fermentation. If the oxygen requirement per tonne of substrate is taken as 1 for molasses, then n-paraffins need 2·5 and methane 5 times as much oxygen. Thus a cheap substrate may have high energy costs due to oxygen transfer requirements.

Fertilizer. A chemical which promotes plant growth by enhancing the supply of essential nutrients such as nitrogen, phosphate and potash. Fertilizers can be inorganic, such as ammonium sulphate (NH_3SO_4) or lime, or organic, such as sewage SLUDGE or manure. They can be added to the soil or, at low concentrations, sprayed on foliage as a foliage feed.

Fertilizer run-off can be a major threat to watercourses and has already caused enrichment in the Great Lakes (◊EUTROPHICATION).

Where natural fixation of nitrogen is taking place (◊NITROGEN CYCLE), as in a clover crop, the addition of inorganic nitrogenous fertilizers can inhibit the natural fixation.

Fixation of nitrogen by man for fertilizers now equals the natural fixation rate, and the attendant increase in NITROGEN OXIDES may be a threat to the ozone shield.

Plant breeders are now attempting to develop plants which will fix nitrogen directly from the atmosphere, and so eliminate or decrease the need for nitrogenous fertilizers.

Although some agronomists believe that further increases in farm production are possible by increased use of fertilizers, some leading agricultural chemists are convinced that we may be close to the economic limit of fertilizer use and that, with the exception of grassland, the economic limit may already have been exceeded. Furthermore, there is reason to suspect that artificial fertilizers alter the soil ecology in a detrimental manner. Such effects are cumulative, and farmers often find that, with time, they are using increasing amounts of fertilizer to achieve the same yield. (◊SOIL, FERTILITY AND EROSION OF; MODERN FARMING METHODS; METHAEMOGLOBINAEMIA.)

Fertilizer supplies. Britain imports most of its artificial fertilizers from abroad, either as finished fertilizer or in the form of raw materials. With the exception of phosphates, this situation will probably change when North Sea oil becomes more available and as the high-quality potash deposits beneath the North York Moors are developed.

The basic raw material of nitrogen fertilizers (e.g. ammonium sulphate, ammonium nitrate) is ammonia (NH_3), manufactured from atmospheric nitrogen and hydrogen obtained from hydrocarbons such as methane or naphtha (a petroleum derivative). The price of nitrogen fertilizers has recently risen in line with the increasing price of oil, and developing and UNDERDEVELOPED COUNTRIES are having great difficulty in maintaining their essential supplies without external aid.

Two-thirds of the world's known deposits of calcium phosphate are in Morocco, although there is a possibility that the sea-bed may yield further supplies. (◊ SEA, MINING OF.)

Fibrosis. A scarring of the lung tissues, caused by dust inhalation. Almost all dust diseases, such as pneumoconiosis (coal-dust disease) and silicosis (stone-dust disease caused by mining, quarrying, shot blasting and stone-dressing operations) are characterized by a scarring of the lungs.

Other dusts which have been implicated in lung disorders include talc, fireclay, mica, china clay, graphite and ASBESTOS. Flax and hemp dust give rise to a disabling lung disease called byssinosis.

Field moisture capacity. The equilibrium amount of water retained in the SOIL after excess water has drained away. In the UK the soil generally reaches the field moisture capacity in winter and early spring. Thereafter, evaporation removes moisture and causes a soil moisture deficit.

First law of thermodynamics. ◊ LAWS OF THERMODYNAMICS.

Fission. The spontaneous or induced splitting of heavy atoms (uranium, plutonium) into two roughly equal parts, thereby releasing large quantities of energy. (◊ NUCLEAR ENERGY; NUCLEAR REACTORS, CLASSES; NUCLEAR REACTOR DESIGNS.)

Fixation. ◊ NITROGEN CYCLE.

Fixed carbon. A measure of the primary productivity of an ecosystem based on the amount of carbon fixed by PHOTOSYNTHESIS per unit area. (◊ CARBON CYCLE.)

Flash point. The temperature at which an inflammable liquid gives off sufficient vapour to catch fire when ignited. (◊ VOLATILIZATION.)

Fluidized bed. A form of solid fuel or refuse combustion in which the fire bed is fluidized by the combustion air blown upwards through it. This produces highly efficient combustion and allows the use of low-grade fuels that are not suitable for conventional combustion plant designs.

Fluidized beds can also be used as separation devices, e.g. a bed of sand fluidized by air can be used to separate metals from glass or plastics. The lighter components float, the dense components sink.

Fluidized beds also lead to the desulphurization of fuels by using a suitable material, e.g. powdered limestone, to combine with the sulphur and prevent the emission of sulphur oxides.

Fluoridation. The addition of a fluoride salt in trace quantities to public drinking water for improving resistance to dental caries. The aim is to adjust the fluorine content to the optimum amount of 1 part per million. Many studies have shown that when a population is supplied with water containing such a concentration the incidence of dental decay is at a minimum in the population.

Fluorine (F). A greenish-yellow gas, and the most reactive element known. It is highly poisonous. Fluorine is used in the manufacture of halogenated fluorocarbons or CHLOROFLUOROMETHANES used in AEROSOL PROPELLANTS.

Fluorides are released into the environment by brick factories, aluminium smelting and phosphate works. The fluoride particles can be deposited on the herbage in the vicinity of process plants, and there are many recorded cases of cattle suffering fluorosis which leads to loss of teeth and bone growths at joints, giving rise to lameness. Fluorides can also affect plants by entering the stomata and then moving to leaf margins where they accumulate ($\diamond$GLADIOLI). Extremely small concentrations, as low as 0·005 part per million, will blight maize, and 0·001 part per million lowers citrus productivity (US data).

The release of hydrogen fluoride in concentrations in excess of 1 microgramme per cubic metre ($1\ \mu g\ m^{-3}$) can lead to herbage fluoride levels in excess of 30 to 35 parts per million, at which level dairy cattle are at risk. However, the MAXIMUM ALLOWABLE CONCENTRATION based on a three minute mean for fluoride release is $75\ \mu g\ m^{-3}$, which corresponds to long-term mean of $3·7–11\ \mu g\ m^{-3}$. Thus, it is possible that while the discharge regulations are met, the grass in the vicinity of such an operation may not meet the guidelines for prevention of fluorosis in cattle unless special agricultural practices are adopted.

Fluorosis. $\diamond$FLUORINE.

Fly ash. The finely divided particles of ash readily entrained in the flue gases arising from the combustion of fossil fuels (mainly coal). The particles of ash may contain unburnt fuel. ($\diamond$ELECTROSTATIC PRECIPITATOR.)

Fog. Microscopic water droplets, varying from 2 to 20 microns in diameter, suspended in air and reducing visibility. One form, ice fog occurs in regions where the ambient temperature is less than $-20°C$. It is formed by the spontaneous nucleation and freezing of water vapour in combustion gases from power stations or vehicular exhausts in Arctic regions.

Both forms of fog substantially reduce visibility. Ice fog has been known to cut visibility down to less than 2 metres, as can an extremely dense 'pea-souper'.

In urban areas the density of fog is closely related to the amount of particulate material present in the air and upon which moisture can condense easily as droplets.

Food additives. Substances used to treat food during processing. They include colouring materials, bleaches, anti-oxidants, food extenders, sweeteners, tenderizers, flavours and nutritional supplements. They are carefully controlled by legislation, which is continually being revised. The extent of the control is very varied in different countries. Legislation in the UK is based on the Food and Drugs Act 1955.

Food chain. A series of organisms through which energy is transferred. Each link feeds on the one before it (except for the first one which is herbage – see Figure 30) and is eaten by the one following it. Herbage is said to belong to the class known as producers, which transform solar energy and carbon dioxide into sugars via photosynthesis. The remaining links are consumers. Consumers are ordered first, second, and so on. Thus a three-stage food chain would be grass consumed by cattle consumed by man, i.e. producer, first-order consumer, second-order consumer. A food chain is essentially an energy conversion scheme and as in any energy conversion device there are conversion efficiencies and transfer losses. Hence, the energy fixed by producers will always be greater than that fixed by first-order consumers, which in turn will always be greater than that fixed by second-order consumers, and so on (Figure 30).

Those organisms whose food is obtained from green plants and have the same number of links in their chain are said to occupy the same trophic or energy level. Thus a cow and a rabbit are both at the same trophic level. The shorter the chain, the more food there is available, e.g. 5 kilogrammes (5 kg) grain may at best produce 1 kg live-weight gain on cattle, and 10 kg meat may produce only 1 kg gain in man. The feed–conversion efficiency is usually very much less than this. If a link can be cut out of the chain, e.g. by man consuming grain directly (he may still need a protein supplement), then 5 kg grain may produce 1 kg in man instead of 0·1 kg as formerly. ($\Diamond$ FOOD WEB.)

Food from the sea. Perhaps one of the most widely promoted of the ecological myths is the one which asserts the existence of an almost limitless food supply in our seas which is there for the taking.

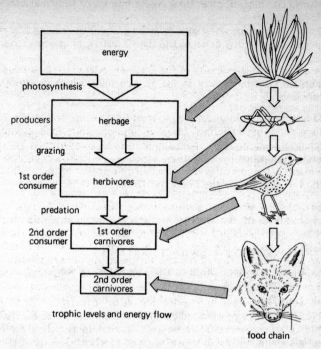

Figure 30. Trophic levels and energy flow, and an example of an associated food chain.

The marine biologist J. H. Ryther provides a more realistic perpective:

The open sea – 90% of the ocean and nearly three-fourths of the earth's surface – is essentially a biological desert. It produces a negligible fraction of the world's fish catch at present and has little or no potential for yielding more in the future.

High productivity of edible sea-life is limited almost exclusively to those off-shore and coastal areas where estuaries and powerful upwelling currents bring nutrients to the surface and where the sea is shallow enough to permit the creation of phytoplankton by photosynthesis of sunlight. These areas represent the world's principal fishing grounds; they are also the areas most affected by coastal and estuary pollution which is increasing yearly.

The following table summarizes the world situation:

Area	Percentage of ocean	Area (square kilometres)	Annual fish production (tonnes)
Open ocean	90	326 000 000	160 000
Coastal zone[a]	9·9	36 000 000	120 000 000
Coastal up-welling areas	0·1	360 000	120 000 000
	Total annual fish production		240 160 000
	Amount available for sustained harvesting[b]		100 000 000

[a] Including certain offshore areas where hydrographic features bring nutrients to the surface.

[b] Not all the fishes can be taken; many must be left to reproduce or the fishery will be overexploited. Other predators, such as sea birds, also compete with us for the yield.

SOURCE: After J. H. Ryther, 'Photosynthesis and fish production in the sea', *Science*, vol. 166, no. 3901, 3 October 1969, pp. 72–6.

(◊ANCHOVY FISHERIES; WHALE HARVESTS.)

For a detailed assessment of the current state of the UK fishing industry, see *Fish Industry Review*, vol. 4, no. 1, 1974.

Food web. A system of interlocking FOOD CHAINS. A food web comprises all the separate food chains in a community (see Figure 31), including DECOMPOSERS whose role is vital in recycling essential materials and nutrients. (◊CARBON CYCLE; NITROGEN CYCLE.)

Forests, Tropical, Clearing of. ◊PLANT NUTRITION.

Fractionation (green crop). ◊LEAF PROTEIN EXTRACTION.

Freezing. ◊DESALINATION.

Fuel. A source of thermal ENERGY. Fuel can be bacterial, fossil, vegetable or nuclear in origin. (◊CELLULOSE ECONOMY; COAL; ENZYMES; ETHANOL; GAS, NATURAL; HYDROGEN; METHANE; METHANOL; NUCLEAR ENERGY; OIL; OIL SHALES; TAR SANDS.)

Fuel cells. One of a class of devices known as DIRECT ENERGY CONVERTERS. The fuel cell directly converts chemical energy to electrical energy without the intermediate step of random molecular energy, e.g. the raising of steam which needs a turbine and electrical generator before electricity is available. Thus the fuel cell is *not* subject to the Carnot efficiency restriction of the second law of thermodynamics which means that conversion efficiencies of 90 per cent and

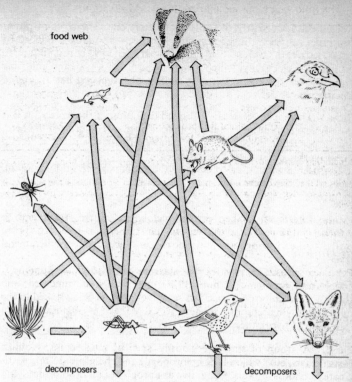

food web

decomposers   decomposers

Figure 31.   Energy flows (food web) in nature.

greater can in theory be obtained ($\diamond$LAWS OF THERMODYNAMICS). The fuel cell requires hydrogen or a hydrocarbon fuel and oxygen from the air. In a simple hydrogen/oxygen fuel cell, there is an electrolyte and two non-consumable electrodes which catalyse the ionizing reactions. Thus at the anode, or positive electrode, hydrogen (H_2) decomposes to $2H^+ + 2e^-$ (2 hydrogen ions + 2 electrons) and at the cathode, or negative electrode, 2 electrons and 2 hydrogen ions combine with a half-molecule of oxygen (written as $\frac{1}{2}O_2$) to form water,

$$\text{i.e.} \qquad 2e^- + 2H^+ + \tfrac{1}{2}O_2 \rightarrow H_2O.$$

The voltage from the reaction is 1 to 1·5 volts per cell and any voltage can be obtained by connecting fuel cells in series as with ordinary batteries.

Fuel cells using coal or oil have been mooted but so far the hydrogen/oxygen cell is currently the only practicable one. The costs of power from fuel cells is not cheap *and* the hydrogen is usually produced by ELECTROLYSIS which demands primary energy consumption to produce the electricity, which means in turn that a system comprising a nuclear reactor and boiler plus turbine plus fuel cell has an *overall efficiency* of 30 per cent or less.

Fume. There is no generally accepted definition of the word, but it is usually taken to mean minute particles less than 1 micron in diameter and suspended in the air. They are usually released as a result of certain metal working and chemical processes.

The term is often used to describe the vapours given off by a liquid, especially if it is offensive or toxic. (◊ FUMES, EFFECTS OF; VOLATILIZATION.)

Fumes, Effects of. When certain metals are heated, they tend to volatize and fumes (i.e. particles less than 1 micron in size) can spread easily. Those of manganese and zinc can produce an effect called metal fever.

Fumes released in plastics manufacture can also cause numbness, cramps and impotency. Teflon (PTFE) has been so documented; and recently vinyl chloride disease has been identified as being linked to the fumes from vinyl chloride monomer (VCM), as has angiosarcoma, a rare form of liver cancer (◊ POLYVINYL CHLORIDE).

Fungi. A group of simple organisms such as mushrooms, moulds, rusts, yeasts, etc., which lack chlorophyll, and therefore do not participate in photosynthesis. Fungi live as aprophytes on dead or decaying organic material or as parasites on living material.

Fungi have a high production of enzymes and selected varieties can decompose wood rapidly, as in dry rot. Filamentous fungi can hunt for food, can colonize, and can penetrate between cellulose fibres and decompose them. The use of filamentous fungi has been proposed to make high-grade feedstuffs from STRAW.

YEASTS are unicellular fungi of great commercial importance.

Furfural (C₄H₃OCHO). Organic solvent or intermediary which can be obtained by acid HYDROLYSIS of PENTOSANS contained in waste agricultural products to pentoses and thence to furfural. Almost any plant material can be used and the production of furfural from these sources is receiving considerable attention as a means of utilizing 'income' (◊ SOLAR ENERGY) resources. In the USA, over 2 million tonnes could be produced from maize cobs alone. Almost any petrochemical can be made from it including nylon intermediates. (The potential yield from sugar cane alone is 50 million tonnes per year.)

In 1970 approximately 40 million tonnes of primary petrochemicals were manufactured in the USA. Thus, this is a potential route for these products as the costs of the synthetic route from hydrocarbons increases, as they undoubtedly will.

Fusion. ◊NUCLEAR FUSION; NUCLEAR ENERGY; NUCLEAR REACTORS, CLASSES; NUCLEAR REACTOR DESIGNS.

G

Garbage. ◊DOMESTIC REFUSE.

Gas, Natural. Gas that occurs naturally, usually in association with oil reserves (though decreasingly so as gas exploration techniques improve). It is an exceptionally clean and convenient fuel whose use for low-grade purposes (e.g. raising steam) will probably be banned in most countries within a few years for the simple reason that it is too valuable to be burned. The energy content of ultimately recoverable world resources is of the same order for gas as for oil. Cheap reserves of gas may be slightly closer to depletion than those of oil, however, because growth in demand has been more rapid. Gas now provides one-third of energy in the USA and is the sixth largest industry. The USSR apparently holds about one-third of ultimately recoverable world gas resources, and North America and the Middle East about one-fifth.

Gas is far more costly to transport overseas than oil. Great efforts are being made to augment pipelines with liquefied natural gas (LNG) marine carriers, so as to export more gas from surplus to deficit areas and to help eliminate the flaring that still wastes a substantial fraction of the gas.

A. B. Lovins, *World Energy Strategies*, Earth Resources Research for Friends of the Earth, 1973.

Gaseous pollutants, Control. ◊THRESHOLD LIMITING VALUE; THREE MINUTE MEAN CONCENTRATION; EMISSION STANDARD.

Gases, Properties of. Gases are compressible and their density is proportional to pressure (all other things remaining constant). They expand in proportion to temperature and thus the greater the temperature, the lighter or less dense is the gas. Thus a minimum gas exit temperature may be specified in some air pollution discharge consents to acid dispersion.

Genetic erosion. Genetic erosion occurs when many varieties of plant (or animal) are allowed to die out so that they are no longer available for breeding. Modern plant-breeding methods have led to the emergence of hybrid species which crop uniformly, have x number of peas of y millimetres diameter in the pods at the nth day of the mth month and so on. The seeds are obtained by crossing parents which have been selected for certain attributes. This method obtains the right crop at the right moment all ready for harvest in unison, but is often done at the expense of vigour and genetic potential for, say, resistance to insect attack or hardiness to drought or change in temperature. Once this genetic potential is lost, it cannot be recovered. Thus, to prevent this genetic erosion, the retention of the greatest genetic variety possible is essential in seed banks if only as an insurance policy against the day when pesticides lose their efficacy once and for all, or against the various blights that agrochemicals might just not manage to 'cure'.

The foregoing also applies to domestic animals. The necessity for breeding varieties of sheep, cattle, etc., is just as important. The ecological maxim that there is stability in diversity is more apt than ever. (◊ECOSYSTEMS.)

Genetic load. ◊IONIZING RADIATION.

Geothermal power. The use of energy from the earth's interior conducted to the surface in a few areas of the globe. To be useful the energy must be available in superheated water or steam form and it may then be used in a conventional power plant. Italy, USA, Iceland, USSR and New Zealand all have suitable geothermal energy fields. This method of power generation is a strictly local and usually small-scale affair and very useful to those areas where it occurs; as a fraction of global energy requirements it is very small.

The use of so-called 'hot rocks' two to five miles under the earth's crust could also be considered as a free energy source – but at what price to extract the heat? Just as with DISTRICT HEATING, the economic feasibility of such schemes is all important.

Ghost acreage. That acreage that would need to be worked by a given country in order to produce an equivalent nutritional output in the absence of cheap protein imports.

The concept of ghost acreage was introduced by Georg Borgstrom to illustrate the dependence of modern industrial farming on imports from the underdeveloped countries (◊AGRICULTURAL ECONOMICS). Borgstrom divides ghost acreage into fish and trade acreage. Fish acreage is defined as the average tilled land required to raise an amount of animal protein equivalent to that provided by fisheries via food and feeding stuffs, taking into account the present techniques of agriculture in each country. Trade acreage is computed by the same methods, but refers to the net amounts of agricultural products being imported or exported for food or feed.

The table below shows the heavy dependence of the highly industrialized countries on cheap imports from less wealthy nations – an arrangement which is unlikely to continue as populations increase and basic resources diminish. There are already signs, in the huge increase in cost of imported animal feed, that the 'golden years' are over. We shall soon have to make more realistic estimates of the number of people our country can truly feed, and of the real costs involved.

Ghost acreage of selected countries (1963–5) as a percentage of tilled land

	Fish acreage	Trade acreage	Ghost acreage
Japan	214%	120%	334%
UK	65	240	305
Netherlands	165	105	270
West Indies	120	110	230
Israel	74	140	214
West Germany	73	135	208
Italy	19·5	29	48·5

From Georg Borgstrom, 'Ecological aspects of protein feeding – the case of Peru', in M. Taghi Farvar and J. P. Milton (eds.), *The Careless Technology*, Tom Stacey Ltd, 1973.

Gladioli. Flowers which are useful indicators of the presence and concentration of airborne FLUORIDES. The leaves mottle and turn yellow-brown in the presence of concentrations as low as 0·5 microgrammes per cubic metre. Skilled interpretation is of course required. The variety 'Snow Princess' is most commonly used as the most sensitive, while other varieties are progressively more resistant to fluoride damage.

Glucose. Often called the key sugar as it is important to both plants and animals as an energy-producer. It is a monosaccharide hexose ($C_6H_{12}O_6$) and is present naturally in fruits. It also results from the hydrolysis of other sugars and starches such as sucrose, lactose, maltose, cellulose and glycogen. An example of this in commercial use is the enzymic conversion of maltose and sucrose to glucose by yeast in the brewing process. (◊SUGARS; ENZYME TECHNOLOGY.)

Granuloma. Multiple accumulation of cells distributed in nodules in the lung. It is not particularly common. Beryllium, tungsten carbide and zinc dusts have been responsible.

See G. L. Waldbott, *Health Effects of Environmental Pollutants*, C. V. Mosby, St Louis, 1973.

Greenhouse effect. The mechanism whereby incoming solar radiation is trapped by a glass sheet or blanket of CARBON DIOXIDE; as both are transparent to solar radiation, the short-wave incoming radiation is transmitted. However, both substances are opaque to long-wave re-radiation from the earth's surface or any other object underneath. Thus, the heat is trapped and the underlying surface is thereby warmed and, in a greenhouse or in SOLAR HEATING, the heat may be put to use.

Fears have been expressed that the earth's surface temperature could rise due to the build up of carbon dioxide in the atmosphere and the attendant solar heat gain could disrupt climate patterns. (◊ATMO-SPHERE.)

Green Revolution, The. The most widely recommended means of increasing agricultural yields is through the increased use of FERTI-LIZERS and the introduction of new 'high yield' varieties of grain.

Fertilizers are easily produced (although the ENERGY required in their production is considerable) and have been used intensively for many years now. However, the environmental consequences of the intensive use of fertilizers and the effects of a really large-scale increase in their use are incalculable. In addition, the difficulties of implementing the proposed increase in fertilizer use on the scale required are immense. Ehrlich has calculated that, if India was to apply fertilizer at the *per capita* level employed by the Netherlands, India's fertilizer needs alone would amount to nearly half the present world output.

The second proposal for increasing yields is to develop new high-yield or high-protein strains of food crops. Such new strains have had considerable success over the past few years, particularly in Asia. They mature early and are relatively insensitive to the length of day, making the production of two or three crops a year possible. But these new strains usually require high fertilizer inputs in order to realize their full potential, and we have already pointed out the problems involved there. Vast amounts of capital are required for fertilizer production and distribution. Abundant water is also necessary, as are PESTICIDES and mechanical planting and harvesting machinery. It is just those countries that need these crops most who are desperately short of capital. Furthermore, we are unsure how resistant these new strains are to the attacks of insects and plant diseases.

Therefore, although high-yield agriculture is promising, it is unlikely that it will ever fulfil its promise. Lack of capital, expertise, and most of all, lack of time and the will to act quickly will probably mean that at best the Green Revolution will only allow us to keep pace with population growth for a couple of decades. (◊GHOST ACREAGE.)

P. R. Ehrlich and A. H. Ehrlich, *Population, Resources, Environment*, W. H. Freeman, 1970.

Gross national product (GNP). A measure of the total flow of goods and services produced by the economy over a particular period – usually a year. It is obtained by adding up, at market prices, the total national output of goods and services. 'Intermediate' products are not included since it is assumed that their value is implicitly included in the prices of the final goods. To this total figure (frequently termed the gross domestic product) is added any income accruing to residents arising from investment abroad, and from it is deducted any income earned *in* the domestic market by foreigners abroad. The final figure is called the gross national product and it is generally supposed to be a measure of economic success.

However, as a realistic guide to a nation's economic well-being there is a lot wrong with the GNP. First, it includes the very considerable expenditure on arms and the military, which is totally non-productive. Second, it includes such things as pollution. When somebody pollutes the environment and somebody else cleans it up, the cleaning-up process is included in the GNP. Similarly, if you are an urban dweller whose health is affected by the pollution, then your hospital bills also contribute to the GNP. But the main defect with the GNP as an indicator of economic well-being is its preoccupation with indiscriminate production (◊ECONOMIC GROWTH).

For a useful summary of the problem, see K. Boulding, 'Fun and games with the gross national product', in H. W. Helfrich (ed.), *The Environmental Crisis*, Yale University Press, 1970, pp. 157–70.

Ground-level concentration. The concentration (amount per unit volume) of a pollutant in air between ground and about 2 metres above the ground, i.e. at breathing level.

H

Haber process. ⟡NITROGEN CYCLE.

Haemolysis tests. Haemolysis is the breaking of blood corpuscles by the action of a poisonous substance. It may be used as a means of testing for the biological activity of a suspected substance by incubating the substance with red blood cells, and then measuring the amount of haemoglobin released by the breaking of cells.

Initial haemolysis tests on PVC powder suggest that PVC dust is as biologically active as blue asbestos or crocidolite (⟡ASBESTOS). *If* this is the case, it has serious ramifications concerning the handling of PVC powders which are used in many plastics plants.

See L. McGinty, 'Feedback', *New Scientist*, 26 June 1975, p. 714.

Half-life. The time taken for half the quantity of a substance to disappear from the environment, e.g. by biodegradation or by discharge from a biological system. For example, inorganic MERCURY has a half-life of six days in humans; organic mercury compounds have a half-life of 70 days. Thus, the ingestion of mercury in food will result in totally different body burdens depending on the form (see Figure 32).

Radioactive half-life is the time taken for half the number of atoms of a radioactive substance to decay to atoms of a different mass. Different RADIONUCLIDES have different half-lives: plutonium–239 has a half-life of 24400 years; strontium–90 has a half-life of 28 years; xenon–138 has a half-life of 17 minutes. (⟡EXPONENTIAL CURVE.)

Radiological half-life: in estimating the dose of radiation from radioactive matter that has been ingested, both the radioactivity of the matter and the residence time in the body are important and give rise to the concept of radiological half-life.

Various organs have different half-lives for the presence of the same compound or element. For example, an ingestion of organic mercury may have a half-life of 50 days in the liver, but 150 days in the brain (see Figure 33). Thus, even though the brain will get a much smaller proportion of the input, the bulk going to the kidneys, at the end of

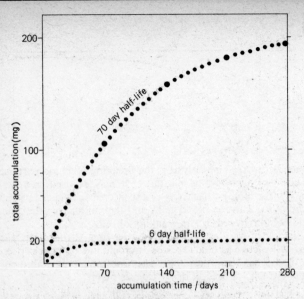

Figure 32. The effect of different periods of half-life on accumulation in the body. Organic mercury reaches ten times the level of inorganic mercury in nine months. At the 2 milligramme per day ingestion level shown here symptoms of severe poisoning from organic mercury would in fact appear before the third month. (From A. Tucker, *The Toxic Metals*, Pan; Ballantine, 1972.)

approximately a year the brain concentration is higher. This accounts for the effect of heavy metal poisoning on the central nervous system as at Minamata (◊MINAMATA DISEASE).

A. Tucker, *The Toxic Metals*, Pan/Ballantine; Earth Island, 1972.

Halogenated fluorocarbons. Group name for compounds, such as ethane or methane, in which some or all of the hydrogen in their structures is replaced by chlorine and/or fluorine. They are used as refrigerants and aerosol propellants because of their low boiling point. An example of a refrigerant is Freon–12 (CCl_2F_2), which has a boiling point of 28°C. TRICHLOROFLUORMETHANE (CCl_3F) is used as an aerosol propellant.

Halogenation. Reaction with a member of the halogen group: namely, fluorine, chlorine, bromine, iodine.

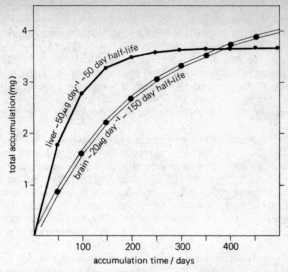

Figure 33. Accumulation in brain and liver assuming 20 per cent and 50 per cent absorption respectively of a daily dose of 100 microgrammes – and a longer brain half-life. A dose of 100 microgrammes would be contained in a normal single meal of fish that contained 0·5 ppm mercury (the permitted maximum in the USA and Canada). Brain half-life is not known accurately but is believed to be considerably longer than that of the liver. Half-life in the testes is also long. According to some scientists, symptoms of damage to the brain first appear at concentrations as low as 3 microgrammes per gramme of brain tissue – which would be reached in 50 days on this half-life basis. Other scientists believe that 20 microgrammes per gramme of brain tissue is the damage-symptom threshold. (From A. Tucker, *The Toxic Metals*, Pan; Ballantine, 1972.)

Hazardous pollutants. This is a classification used by the United States Environmental Protection Agency. A hazardous pollutant is one to which even slight exposures may cause serious illness or death. MERCURY, ASBESTOS and BERYLLIUM have all been so classified.

Haze. Dust, salt or smoke particles often suspended in water droplets in suspension in the atmosphere. Haze can also be caused by substances called terpenes given off naturally by vegetation and forests.

Hearing. The ear responds in varying degrees to sound of different frequencies. The range of frequencies it can encompass is roughly from 20 to 20 000 hertz (Hz).

For a given sound pressure level a high frequency sound appears to be stronger than a low frequency one. The loudness of sounds as judged by the ear depends on sound pressure level and frequency. Thus the measurement of sound is weighted in a similar way to the frequency response of the ear and the standard weighting used is called the A-scale. Meter readings on such a weighting are called sound levels measured in decibels on the A-scale or dB(A) (◊DECIBELS A-SCALE).

Measurements of sound level in dB(A) are widely used and have been found in most cases to be in agreement with typical assessments of loudness. The level in a bedroom with the windows open in a quiet suburban area at night would be typically 35 dB(A). That in a large shop would be about 60 dB(A), and in a busy workshop where raised voices have to be used for conversation it would be about 85 dB(A). A change of loudness is just noticeable in normal circumstances when the sound level changes by 3 dB(A). A doubling or halving of loudness is apparent when the level rises or falls by 10 dB(A) (◊DECIBEL).

There is only one important source for which dB(A) are not normally used, and that is aircraft. The sound generated by aircraft engines has predominant components in particular frequency bands, and these can have a significant *subjective* effect for which A-weighted sound levels do not adequately account. Measurements are therefore made (in decibels) in each of a number of restricted frequency bands: from these a total level is calculated, giving due emphasis to the predominant components. The calculated total is called the 'perceived noise level', the units of which are PNdB. The precise numerical difference between PNdB and dB(A) varies depending on which frequency bands contain the predominant sound. The bands differ with different types of aircraft engine. However, values of PNdB are higher than values of dB(A) would be for the same sound, and as a rough guide a difference of 13 can be assumed. Certain types of new aircraft are now subject to noise certification before they come into service. For this purpose the perceived noise level is adjusted to take account of any pure tones (single-frequency notes) in the noise, and of the length of time for which the higher noise levels are experienced. The result is termed the 'effective perceived noise level', the units of which are EPNdB. (◊NOISE; SOUND; ROAD TRAFFIC NOISE; AIRCRAFT NOISE; INDUSTRIAL NOISE MEASUREMENT; NOISE INDICES.)

Heat island. ◊TEMPERATURE INVERSION.

Heat pump. A device for pumping heat from a low temperature source, e.g. the surrounding ground or a stream, and delivering it at a higher temperature; in other words, a refrigerator in reverse. Heat pumps require a mechanical or electrical energy input and therefore the performance coefficient or ratio of heat produced to electricity consumed is very important. As explained under the second law of thermodynamics (⟡LAWS OF THERMODYNAMICS), the conversion of heat to electricity in most power stations is about 3 units heat for one of electricity, so to obtain wide applicability heat pumps must have coefficients of performance (c.o.p.) greater than 3, otherwise they would consume a greater heat equivalent in electricity than they were producing. Theoretically, heat pumps can have c.o.p.s of 8–10, but practically these are much less.

Heavy metals. MERCURY, LEAD, CADMIUM, SELENIUM, CHROMIUM and PLUTONIUM are among the so-called heavy metals – those with a high atomic weight. (Note that some are not metals at all. The term is rather loose and is taken by some to include arsenic, BERYLLIUM, manganese and nickel, in addition to those listed above.) Some of the above, if ingested in minute quantities, can be lethal (1 or 2 microgrammes for some; in the case of plutonium very much less). With the exception of plutonium, the others are in common use in industrial processes and therefore they may be discharged to the environment to be mobilized by the action of air and water and concentrated by the action of living organisms, and, as in the classic case of MINAMATA DISEASE, can cause death and deformity.

There are no genetic mechanisms for protection from heavy metals and because of their tendency to accumulate in selective organs such as the brain or liver, average intake estimates are of little use. The dose to each individual organ must be considered separately as is the case with IONIZING RADIATION, because some parts of the body are more vulnerable than others (⟡HALF-LIFE).

Ecosystems can be affected by their discharge, e.g. marine and freshwater diatoms can lose 50 per cent of their growth rate at concentrations as low as 1 part per thousand million (10^9) of mercuric compounds. Because of their environmental importance, each is discussed individually.

See A. Tucker, *The Toxic Metals*, Pan/Ballantine; Earth Island, 1972.

Hepatitis. A disease that can be caught by drinking sewage-contaminated water, and which causes liver inflammation. (⟡PUBLIC HEALTH.)

Herbicides; also known as defoliants. Chemicals used for the elimination of unwanted plants or the total elimination of all plant growth.

They are extensively used in agriculture ranging from the selective control of couch grass or wild oats to ground clearance of overgrown areas. Other uses are in roadside maintenance, railway tracks clearance, etc. There are two groups: one is the 2,4-D; 2,4,5-T set (chlorophenoxy acid herbicides), which causes changes in plant metabolism; the other set (symmetrical triazines), simazin, diruon, interfere with photosynthesis and the plants starve to death. Both sets have minimal direct effect on animals *but* they have enormous ecological implications by destroying food sources, habitats, etc. Their use is yet another example of man changing ECOSYSTEMS to suit himself. The use of defoliants, as in Vietnam or in jungle clearance for agriculture, can permanently destroy tropical forests. Once the tree cover is removed, the soil is subjected to EROSION and precious nutrients are rapidly leached away (◊LEACHING). The indirect effects are that congenital malformations can take place and 2,4,5-T is teratogenic (◊TERATOGEN) due to an impurity called dioxin.

The United States Environmental Protection Agency has now banned most crop-related uses of 2,4,5-T; its use in rangelands has been queried and the ruling is that animals due for slaughter must not be grazed on treated areas within two weeks of slaughter.

Herbivore.   A plant-eater or first-order consumer. (◊FOOD CHAIN.)

Herring catch.   ◊MAXIMUM SUSTAINABLE YIELD.

hertz (Hz).   SI unit of frequency, i.e. number of occurrences or cycles per second, as used in the frequency of radio waves, sounds, vibrations, etc.

Heterotrophic organism.   An organism that requires organic material (food) from the environment, i.e. all animals, fungi, yeasts, and most bacteria are heterotrophic. For their food supply these organisms eventually rely on the activities of the AUTOTROPHIC ORGANISMS.

Hexa-chrome.   ◊CHROME WASTE.

Hydroelectricity.   ◊WATER POWER.

Hydrogen (H).   Colourless, odourless gas, normally existing as the molecule H_2. It is the lightest substance in existence, atomic weight 1. It is extremely flammable and when burned combines with oxygen to form water with heat liberated in the process.

It can be readily converted to other *liquid* fuels, e.g. methanol, ammonia, hydrazine, and can of course be pumped readily by pipeline. These attributes have made many people propose the 'hydrogen fuel economy' as an alternative source of energy when conventional ENERGY resources are no longer available. The hydrogen would be manufactured electrolytically (◊ELECTROLYSIS) or by hydrogenation of remaining coal supplies. The hydrogen fuel economy is based on the premise that large quantities of NUCLEAR ENERGY will be available for the generation of the necessary power for the electrolysis.

See W. E. Winsche, K. C. Hoffman and F. J. Salzano, 'Hydrogen: its future role in the nation's energy economy', *Science*, vol. 180, ro. 4093, 29 June 1973, pp. 1325–32.

Hydrogen sulphide (H₂S). Dense, colourless gas with a smell of rotten eggs. It is extremely toxic. It is produced under anaerobic decay conditions and can accumulate in sewers. This has accounted for several fatalities. It is also produced in industrial processes such as oil refining, chemical manufacture and wood pulp processing. (◊TOXIC WASTES.)

Hydrological cycle. The means by which water is circulated in the biosphere. Evapo-transpiration (◊TRANSPIRATION) from the land mass plus evaporation from the oceans is counterbalanced by cooling in the atmosphere and precipitation over both land and oceans (see Figure 34). The hydrological cycle requires that on a world-wide basis the evaporation and precipitation are equal. However, oceanic evaporation is greater than oceanic precipitation, thus an excess precipitation is

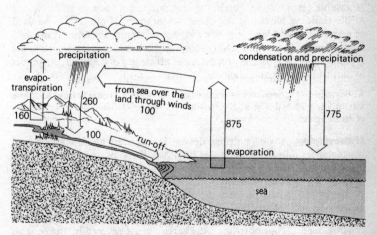

Figure 34. Hydrological cycle in cubic kilometres per day. (Data from G. Borgstrom, *Too Many*, Macmillan Co., 1967.)

133

given to the land. Eventually this land precipitation ends up in lakes and rivers and thus eventually returns to the sea, so completing the CYCLE.

Man intercepts the land precipitation by means of RESERVOIRS or river abstraction, and so obtains his water supplies, but after use by man this abstracted water still ends up in the sea – its arrival there is merely delayed.

Water is also required for PHOTOSYNTHESIS but the fraction of water so used compared with that which is transpired by green plants is less than 1 per cent.

Hydrology. The science concerned with the occurrence and circulation of water in all its phases and modes and the relationship of these to man.

Hydrolysis. The breaking down of a substance by interaction with water. The hydrolysis of carbohydrates, e.g. conversion of starches and sugars, is of major importance in the food and brewing industries. Carbohydrates may be hydrolysed biologically with the aid of ENZYME TECHNOLOGY or chemically at high temperatures with an acid or alkali present as a catalyst. The hydrolysis of CELLULOSE produces fermentable sugars which can be processed by fermentation to yield ethyl alcohol, butanol, acetic acid or PROTEIN. Hydrolysis has been proposed as a refuse-disposal process and is a means of using renewable resources to obtain organic molecules such as ethanol, one of the building blocks of the chemical industry. It could also be used as a source of ethylene, e.g. for PVC manufacture. Protein production is economically feasible. Laboratory work is proceeding in both the UK and the USA on the processing of all classes of cellulosic materials by both biological and chemical means.

A. Porteous, 'The recovery of fermentation and other products from cellulose wastes via acid hydrolysis', Symposium on Reuse and Recycling, University of Birmingham, 5 April 1975.

Hydropulper. A wet pulping device for converting dry pulps and waste papers into a fibrous slurry by the addition of water. It is used in the recycling of PAPER and is the basis of a recycling process for DOMESTIC REFUSE. The pulp recovered from domestic refuse is of a very low quality and so far has little commercial value except for low-grade uses such as for felt.

Hydrosphere. That portion of the earth's crust covered by the oceans, seas and lakes. When the complete earth's crust is meant it is often referred to as the hydro-lithosphere. (◊LITHOSPHERE.)

Hyperactivity (associated with behavioural disturbance). Over-activeness, especially in young children and youths. Many experts believe LEAD poisoning to be a major cause of this syndrome in many cases and recommend 'de-leading' as a cure. (⇨CHELATING AGENTS.)

Incineration. A disposal process which relies on combustion to substantially reduce the volume and/or mass of material to be disposed of. Incineration is commonly used in DOMESTIC REFUSE disposal processes where it has three distinct phases: drying of refuse, volatilization, combustion proper. Grit and FUME emission must be controlled and ELECTROSTATIC PRECIPITATORS are commonly employed. If toxic chemicals are being disposed of, water washing by means of scrubbers is often used so that toxic fumes are eliminated.

The heat released in incineration can be put to use for steam raising, for power generation or DISTRICT HEATING. So far such schemes have not been financially successful in the UK, although as a recycling measure the idea has merit in that energy is recovered.

Industrial noise measurement. The measurement used in the UK for industrial noise is the corrected noise level (CNL). This is based on a measure in dB(A) (◇DECIBELS A-SCALE) to which a specified correction is added if a definite continuous tone (whine, hum, etc.) is present. Another correction is added if there are any impulsive irregularities (bangs, clanks, etc.); and if the noise is not continuous, a further correction is applied which depends on the proportion of time for which the bursts of noise last.

Some values of CNL and the situations in which they are found are listed below.

CNL (dB(A))	Situation of measuring point
85	At 30 feet from a building housing an air/steam drop forge hammer
75	At 30 feet from an air compressor housed in a building with louvred doors
60	At 50 feet from a can-making factory. Continuous noise from stamping machinery and from handling of thin sheet metal

Infections, Water-borne. Water-borne infections include cholera, the enteric (intestinal) diseases, TYPHOID and paratyphoid, the viruses causing poliomyelitis and viral hepatitis, the protozoan responsible for amoebic dysentry, Weil's disease, which is spread by infected domestic and wild animals, including rats, and SCHISTOSOMIASIS (Bilharzia) which is associated with irrigation ditches in developing countries (⬦DAM PROJECTS).

PUBLIC HEALTH practice is intended to prevent, among other things, the spread of disease. Thus, drinking water and sewage are kept separate. However, in less sanitary countries, or even in a crisis in more developed countries, if life-support systems break down, water withdrawal from some sources must cease immediately.

Infiltration. The downward movement of water through the soil. The term is also used to denote the contamination of wells by salt water or the spoiling of an aquifer by pollutants.

Infrasound. Low-frequency sound which usually emanates from low revving ship or heavy transport engines in frequencies below 100 Hz. A typical living-room will resonate around 12 Hz thus amplifying heavy traffic noise. One noise measurement in London recorded a value 85 DECIBELS at 5 Hz coming from ships in the docks. Infrasound is not catered for on the dB(A) scale which is weighted to represent the response of the human ear and therefore takes little account of noise below 100 Hz.

The only way of cutting down infrasound is to stop it at source as insulation is rather ineffective in this frequency range.

Insecticides, Synthetic. ⬦CHLORINATED HYDROCARBONS; ORGAN-OPHOSPHATES.

Insects. Insects proper have a head, thorax and abdomen. In this book the word is taken to include all insects plus mites, spiders, ticks, centipedes, etc., and the various stages the insect goes through (metamorphosis) in its life.

Insects are very versatile. They are found everywhere in the BIO-SPHERE. They can go for long periods without food and water. As a class, they have the following attributes for survival:

1. They are extremely adaptable and can live on almost anything of organic origin under conditions totally hostile to man.

2. They are usually small in size and their food and water requirements are often minimal.

3. They are mobile and can often fly to and from food sources, and escape their natural enemies.

4. They have breeding characteristics that enable resistant strains to be rapidly generated, e.g. DDT-resistant flies ($\Diamond$ECOSYSTEM).

As they are usually near the base of the ECOLOGICAL PYRAMID their numbers vastly outnumber others above them in the pyramid. Hence, *insects are more likely to survive in an eradication programme than their predators.*

5. *Reproduction*: Mating is carried out at any suitable opportunity and the sperm stored in the female until conditions are right for fertilization. This is a potent means of survival.

6. *Structure*: the adult body shape is designed for great strength and is coated with hard chitin for protection.

7. *Metamorphosis*: many undergo the following stages of development – egg, larva, pupa, adult. At each stage they have different food and environmental requirements. Consequently they are capable of inflicting great damage on man, his animals and crops. However, a vast majority of insects are not only beneficial to man but essential. ($\Diamond$PEST CONTROL.)

Internal combustion engine. The internal combustion engine is characterized by combustion of the fuel inside an enclosure, usually a cylinder, which causes rapid gas expansion which in turn forces a piston to move and do work. The characteristics of an internal combustion engine are that there is a need for a homogeneous easily ignited fuel (petrol or diesel oil) which, on ignition, does work due to the expansion of the gases which are then released to the atmosphere. The cyclic nature of the operation plus poor fuel distribution in the cylinder means that the THERMAL EFFICIENCY is low, combustion is often incomplete and thus unburnt HYDROCARBONS and CARBON MONOXIDE are released to the atmosphere. NITROUS OXIDES are also released at concentrations which depend on the temperature and duration of the combustion flame. Despite this, the internal combustion engine is in mass production and has market dominance. Attempts to replace it will meet with strong opposition due to the capital invested in its manufacture and aftercare. ($\Diamond$EXTERNAL COMBUSTION ENGINE; AUTOMOBILES; AUTOMOBILE EMISSIONS.)

Ion. An electrically charged atom or molecule produced by a loss or gain of electrons. Gases can be ionized by electrical discharge or ionizing radiation.

Ion exchange. The removal of IONS from solution. It can be carried out by the use of a suitable ion-exchange medium which has the power to remove or capture the desired ions from solution. The choice of medium determines the type of ion removed (positive or negative). Once the medium is saturated it must be taken off-stream and *recharged* by passing an alkaline *or* acidic solution through the bed which replaces the captured ions from the recharging solution. Ion exchange has many uses. For example, in water softening, calcium ions are replaced by sodium ions, and the water is thereby softened. The bed is then recharged with a brine solution which replaces the captured calcium ions and the bed is then ready for reuse. Normally one bed is 'on-stream' while the other is 'off-stream' for recharging.

Ion exchange is used in SEWAGE EFFLUENT TREATMENT, water softening, water purification, SOLUTION MINING, metallic effluent treatment, etc.

Ionizing radiation, Dose measurement. Ionizing radiations may be divided into two main groups:

1. *Electromagnetic radiations* (x-rays and gamma rays), which belong to the same family of electromagnetic radiations as visible light and radio waves.

2. *Corpuscular radiations*, some of which – alpha particles, beta particles (electrons) and protons – are electrically charged, whereas others – neutrons – have no electric charge. This distinction between the two groups becomes blurred, however, when their mode of absorption in materials is considered.

Whilst the exact nature of the biological effects of these radiations is not fully understood, they are related to the ionization that the radiations are capable of producing in living tissue. Thus, the biological effects of all ionizing radiations are essentially similar. However, the distribution of damage throughout the body may be very different according to the type, energy and penetrating power of the radiation involved. Alpha particles from radio-isotopes have ranges of only about 0·001–0·007 centimetres in soft tissue and less in bone. Beta particles have ranges in soft tissues of the order of several millimetres i.e. much greater than those of alpha particles in such tissues. X-rays and gamma rays are not stopped by tissues.

The dose of radiation is the amount of energy absorbed per unit mass of material. The unit is the rad, which is 0·01 joules per kilogramme (10^{-2}J kg^{-1}). However, the rad is a physical measure of energy absorbed; 1 rad of alpha radiation is about four times as damaging biologically as 1 rad of x-rays; 1 rad of alpha radiation is therefore said to have a relative biological effectiveness (RBE) of 4, compared to 1 rad from x-rays.

For biological purposes it is useful to compare doses in biological effect rather than physical units, and the unit is the rem. One rem is biologically equivalent to 1 rad of x-rays. One rem is therefore also equivalent to 0·25 rad of alpha rays. As the rad and the rem are very large dose measures, it is usual to talk in millirads (mrad) or millirems (mrem). (⟡ PLUTONIUM, THE HOT-SPOT CONTROVERSY.)

Ionizing radiation, Effects.   Certain radiations in their passage through matter, are capable of causing ionization, i.e. they can 'knock' electrons out of atoms or molecules or create ions either directly or indirectly. The radiations that do this directly are either fast-moving particles (for example, electrons, protons and alpha particles) or electromagnetic rays such as x-rays or gamma rays. (Gamma rays are similar to x-rays but have much shorter wavelengths and higher energies. The difference is more one of name than of physics.) There are numerous sources of naturally occurring ionizing radiation including the cosmic rays that arrive continuously from outer space, but the sources of environmental significance are certain radioactive substances liberated into the environment by man.

Exposure to ionizing radiation can be harmful as the radiation can cause cancers in the living population and genetic changes that may produce heritable defects in future generations. Ionizing radiation causes mutations, i.e. random changes in the structure of DNA, the long molecule that contains the coded genetic information necessary for the development and functioning of the human being. As the mutations are random events, they are almost certainly harmful to some degree.

The outcome of radiation exposure may be that the cell may suffer so much damage that it dies, or that the cell may continue to function but in a modified way. The second outcome can lead to the uncontrolled growth of a colony of cells derived from the affected one. This is a cancer, and the upset is called somatic (bodily) mutation.

For reasons not fully understood, cancer resulting from irradiation appears only after a long delay. Only recently was the incidence of cancer beginning to show up as abnormal among the survivors of the nuclear explosions at Hiroshima and Nagasaki. An abnormal incidence of leukemia (the uncontrolled growth in numbers of white blood cells), which has a much shorter latency period, was in evidence long ago in the studies of the Atomic Bomb Casualty Commission.

Now if the mutation is in a germ cell (sperm or egg), the entire organism arising from that germ cell together with its progeny will be affected. Many mutations are sufficiently severe to prevent their carriers from living, or at least from reproducing and handing on the defect to later generations. Many others, though, have their effects masked if they are inherited from only one parent. Such mutations are recessive. Their effects appear only in about a quarter of the off-spring of parents *both* of whom carry the *mutant* gene.

There are many undesirable *recessive genes* and most of us carry a few. Consequently from time to time a child of normal healthy parents shows a terrible defect such as idiocy or gross malformation. The hidden undesirable genes in a population are often called its *genetic load*. The tendency of ionizing radiation is to increase the genetic load.

Not all mutations have such dramatic effects. They may result merely in a loss of vigour, a susceptibility to disease, or a reduction in life-span. Such genes may spread widely in human populations that are sustained by medical attention.

Ionizing radiation also tends to produce a reduction in the life-span of the animals actually irradiated. The effect is often regarded as premature ageing. The following figures, published in 1957, are possible examples of this premature ageing.

Group	Average age at death
USA population	65·6 years
Physicians with no known contact with radiation	65·7 years
Radiologists	60·5 years

The radiologists referred to in the table were practising their profession over a period in which the dangers of x-rays were far less clearly appreciated than they are today. The age distribution of the control was also different from that of the radiologists which casts some doubt on drawing absolute conclusions from the data. Nowadays much more sensitive x-ray film is used which requires less intense radiation, and the relevant equipment is carefully shielded to protect its users from undesirable radiation.

It is usual to compare levels of man-made radiation to that of the natural background radiation from cosmic rays and other natural sources. The inference is then drawn that as man is currently contributing radiation amounting to around 1 per cent or less of the natural background, we therefore have a lot of scope for increasing nuclear installations before we need become concerned about the level of man-made radiation. However, because the natural background is irreducible, it does not necessarily follow that this sets an allowable scale for man-made radiation emissions. The effects of an increase of radiation are proportionately very small indeed, but applied to a total population

141

can give rise to large numbers of cancer incidences and genetically defective births. For example, the genetic handicap risk (Maryland Academy of Sciences) for a radiation dose of 170 millirem per year is estimated as $4 \cdot 5 \times 10^{-5}$. The number of cases per annum is then (for the US population of 200 million) $4 \cdot 5 \times 10^{-5} \times 200 \times 10^{6}$, and for a birth-rate of 4·6 million per year the absolute numbers of genetically unsound births per 1000 births is therefore

$$\frac{4 \cdot 5 \times 10^{-5} \times 200 \times 10^{6}}{4 \cdot 6 \times 10^{6}} \times 1000$$

or 2 per 1000 births.

It is worth emphasizing that knowledge of the effects of ionizing radiation on mutation rates is slight and growing only slowly. Estimates for human populations are based on readily observed cases, but mutations that give rise to slightly impaired individuals whose characteristics are nevertheless within the ordinary range of variation are an even more unpleasant prospect. The attrition of human capacities and the deterioration of health and vigour could proceed unnoticed for indefinitely long periods.

It is now generally agreed that, at least for the purposes of estimating hazards, any dose of ionizing radiation, no matter how small or how slowly delivered, must be regarded as prospectively harmful, able to induce genetic mutations and cancer and generally to erode the life-span of its recipient.

The costs of increased exposure to radiation (as in any form of pollution) must be balanced against the social benefits to be derived from the activities leading to the increased exposure. The debate is whether these risks are justifiable and whether we have the right to saddle future generations with our long-life radioactive wastes. ($\Diamond$COST–BENEFIT ANALYSIS; NUCLEAR ENERGY; NUCLEAR REACTOR DESIGN; NUCLEAR REACTOR WASTES; IONIZING RADIATION, DOSE MEASUREMENT.)

Ionizing radiation, Maximum permissible dose and dose limit. The maximum permissible dose of ionizing radiation was originally regarded as one that, in the light of the knowledge available at the time, was not expected to cause appreciable bodily injury to any occupationally exposed person at any time during his life. The phrase 'appreciable bodily injury' was defined as 'any bodily injury or effect that a person would regard as being objectionable and/or competent medical authorities would regard as being deleterious to the health and well-being of the individuals' (International Commission on Radiological Protection, 1966).

The ICRP retains the term 'maximum permissible dose'* for the exposure of radiation workers. By monitoring the doses received by individual workers and by controlling their environment it is possible to ensure that the maximum permissible doses are not exceeded, except in the case of accidents. It is not possible, however, to determine the doses received by members of the public. It is therefore necessary to control the sources giving rise to exposure and to assess the average dose received by a specific group or an entire population, both by means of sampling procedures in the environment and, in appropriate cases, checks on the doses received by a few individuals in the group or population involved. The ICRP believes that the term 'maximum permissible dose' is seldom meaningful in relation to the individual members of the public and has recommended that, in their case, the term 'dose limit' should be used.

The current values of the maximum permissible doses and dose limits for specified organs and tissues, as recommended by the ICRP, are given overleaf. The ICRP has recommended that the genetic dose limit for a whole population should not exceed 5 rem for all sources, in addition to the dose from natural background radiation and from medical procedures. The genetic dose should be kept to the minimum, and the additional contribution to the genetic dose from medical procedures should also be kept to the minimum, consistent with medical requirements. The genetic dose limit of 5 rem is based on a mean child-bearing age of 30 years. The limit, as stated by the ICRP, is equivalent to about 0·15 rem per year (International Commission on Radiological Protection, 1966b).

The dose rates are the suggested *maximum* exposures and represent those dose rates that should never be exceeded by any person occupationally exposed to ionizing radiations. The *recommended* maximum average dose to the same workers is one-third of the tabulated value. Any worker receiving the maximum should be removed from exposure until his annual average rate has been reduced to the currently recommended level.

The occupational risks are considered an admissible exchange for the benefits generated by the occupation. For larger sections of the community, further reductions in exposure are recommended to one-tenth of the occupational doses. This implies a maximum of 0·5 rem

* Observe that this term is likely to be misleading and is often used in a misleading way, because it begs the question, 'What is a permissible dose?' As knowledge of the effects of ionizing radiation has increased over the years, the dose described as 'permissible' has been repeatedly revised downwards. The current value is only a small fraction of earlier estimates of what was tolerable. One may reasonably continue to ask whether the 'maximum permissible dose' is permissible in any commonly understood sense of 'permissible'.

Maximum permissible doses and dose limits

Organ or tissue	Maximum permissible dose for adults exposed in the course of their work[1]	Dose limits for members of the public (average for groups of individuals)
Whole body (in case of uniform irradiation), gonads and red bone marrow	5 rem in a year 3 rem in 13 weeks	0·5 rem in a year
Skin, bone and thyroid	30 rem in a year 8 rem in 13 weeks	3 rem in a year[2]
Other single organs	15 rem in a year 4 rem in 13 weeks	1·4 rem in a year
Hands and fore-arms; feet and ankles	75 rem in a year 20 rem in 13 weeks	7·5 rem in a year

[1] For adults who work in the vicinity of controlled areas or who enter these areas occasionally but are not themselves employed on work involving exposure to radiation, the annual dose to the whole body, gonads and red bone marrow should not exceed 1·5 rem. For other organs and tissues and for the hands and forearms, feet and ankles, the maximum permissible doses are the same as the dose limits for members of the public.

[2] For children up to 16 years of age, the dose limit to the thyroid is 1·5 rem in a year.

annually for a member of a large population and an annual dose of 0·17 rem to the average of a large population. In this context a 'large population' was suggested by a Secretary of the Commission to mean no more than one-tenth of the total population, unless a further reduction in prescribed dose limits be adopted.

Ionization. ◊IONIZING RADIATION, EFFECTS.

Irradiation, Uses of. The major potential uses of irradiation in the environment are:

1. *Insect sterilization.* Irradiated males can be released and, as the mating in some species only takes place once, it is possible to reduce populations of pests without recourse to pesticides.

2. *Digestibility aids.* Many cellulosic wastes can be used by ruminants as an energy source especially if the cellulose structure is broken down. Irradiation is one means of doing this.

3. *Food preservation.* The storage life of iced fish on board ship can be

doubled by a dose of 100 krad without any apparent adverse effects. This technique may enable trawlers to use distant fishing grounds, extend the duration of voyages and deliver a more hygienic product to the customer. Food preservation is an area in which major advances are expected as irradiation has not been shown to produce dangerous radioactivity in foods.

4. *Radioactive tracers.* The pathways of pollutants can be traced by releasing radioactive monitors. This has been done in studies of air pollution and effluent dispersal. They are also of great use in following fluid flow and leak detection in industry, the ingestion/digestion of food in animals, monitoring engine wear and many other applications.

5. *Disinfestation of seeds.* The removal of insect pests from stored grains.

6. *Sterilization.* Large doses of radiation will produce a completely sterile product, e.g. pharmaceutical products and material for transplants.

Irrigation. The application of water to arable soils so that plant growth may be initiated and maintained. Irrigation is used not only in many arid and semi-arid lands, but in the U K to rectify SOIL MOISTURE DEFICITS. In the past eastern England could have benefited from irrigation every three years in ten. The practice may be expected to grow in the U K due to the high S M Ds now being experienced.

Irrigation with brackish water over long periods can eventually render soils unfit for crop growth, as plant evapo-transpiration acts as a distillation process and leaves the dissolved solids in the soil where they accumulate. The spread of Middle Eastern deserts has been partially attributed to early irrigation attempts which resulted in salinization of once fertile soils.

Isotope. Atoms of the same ELEMENT which have the same number of protons in their nuclei (◊ NUCLEUS) but differing numbers of neutrons. Isotopes are written as mass number (i.e. number of protons and neutrons) followed by the symbol, e.g. ^{235}U, which is the fissile isotope of uranium and is capable of sustaining chain reactions in a nuclear reactor.

Isotopes of the same element have identical chemical behaviour but differing physical behaviour.

joule (J). International System of Units (SI) unit of work and energy. It is defined as the work done when a force of 1 newton acts through 1 metre. Commonly the megajoule (MJ) is used because it is a more convenient size for the measure of energy supplied. It is 1 million joules.

The measure of power is the WATT, which is 1 joule per second, or 1 J s^{-1}.

There are several forms of ENERGY, e.g. thermal, chemical, electrical, and joule is applicable to all of them. The form of the energy is important, however, because thermal energy and electrical energy are very different in their ability to do work.

K

Keynesian economics. John Maynard Keynes's *General Theory of Employment, Money and Interest* must be rated as one of the most important works in economics. Its major thesis is to the effect that the automatic working of a free market economy does not, of itself, provide full employment. The advocates of Keynesian economics recognize the need for government intervention to secure full employment in a capitalist economy, although Keynes's other major point – that economic behaviour is based upon incomplete and often mistaken beliefs and is frequently revised when expectations fail to be realized – tends to be overlooked or rejected.

The notion that governments should assume responsibility for safeguarding their national economies against recession and for stimulating a revival of economic activity in the event of a contraction has been accepted in capitalist countries more readily since the Second World War than seemed likely in the 1930s. At that time, Keynes's proposals that, in time of slump, economic activity could be revived by government expenditure, even at the expense of running a deficit on the budget, were viewed with hostility by many. Government expenditure was traditionally regarded as wasteful and a balanced budget was the hallmark of responsible government. Moreover, to urge a more free-spending attitude in hard times went against the traditional sense of thrift and prudence. Reacting to these objections, Keynes pointed out that this alleged waste was in fact buying release from the fetters of depression. To expose the irrationality of prevailing attitudes, he wrote:

If the Treasury were to fill old bottles with bank-notes, bury them at suitable depths in disused coal-mines which are then filled up to the surface with town rubbish, and leave it to private enterprise on well-tried principles of laissez-faire to dig the notes up again (the right to do so being obtained, of course, by tendering for leases of the note-bearing territory), there need be no more unemployment and, with the help of the repercussions, the real income of the community, and its capital wealth also, would probably become a good deal greater than it actually is. It would, indeed, be more sensible to build houses and the like, but if there are political and practical difficulties in the way of this, the above would be better than nothing.

The analogy between this expedient and the gold-mines of the real world is complete (p. 129).

Thus Keynes urged that a purely ritual activity, contributing no material benefit to human well-being, could nevertheless serve to keep the wheels of the economy turning at full speed, and so maintain the vital network of cash-flows upon which every member of an industrial state is dependent for his day-to-day security. (◊ECONOMIC GROWTH.)

J. M. Keynes, *The General Theory of Employment, Money and Interest*, Macmillan, 1961. (First published 1936.)
A. Leijonhufvud, *Keynesian Economics and the Economics of Keynes*, Oxford University, 1968.
For a view of the relationship between economics and politics, sociology and psychology, see G. L. S. Shackle, *Epistemics and Economics*, Cambridge University Press, 1972.

kilocalorie (K). One thousand CALORIES. A unit used in nutritional terms (where it is usually called a 'calorie') to indicate the useful energy

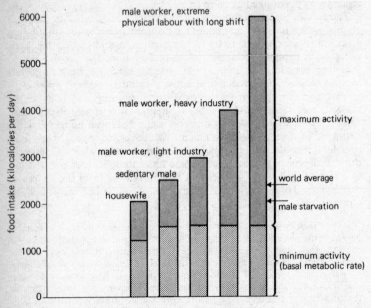

Figure 35. Some typical average caloric needs.

contained in food. Most foods contain a mixture of carbohydrates, fat and PROTEIN, all of which contribute to the energy requirements of the human body. Food needs measured purely in terms of kilocalories are an inadequate way of presenting nutritional data. Information on the availability of protein and other nutrients must also be given. For a table showing the energy value of certain foods, ⬦NUTRITIONAL REQUIREMENTS, HUMAN.

To give an idea of the amount of energy required for certain common tasks performed by an average person, it takes 20 K per hour just to stand still; 215 K per hour to sustain a fast walk; and 1000 K per hour to walk upstairs (or operate a treadmill).

Obviously individual needs for both kilocalories and proteins vary according to body size and activity (see Figure 35), but current world average kilocalorie needs are estimated at around 2350 per capita per day. The United Nations Food and Agricultural Organization estimates a current availability of 2420 kilocalories per capita per day. Taking into account the grossly uneven distribution of food and an average wastage of 10 per cent before consumption, the realities of a predominantly hungry world become evident.

kilowatt hour (kW h). Unit of electrical energy equal to 1000 watt hours or 3·6 megajoules. The kilowatt hour can be used as a measure of ENERGY input in energy analysis. When energy is in its primary form it has the designation $kW\,h_{th}$ which, depending on the efficiency of the conversion, equals 0·25 to 0·4 $kW\,h_e$. (⬦WATT.)

L

Laterite. ⊳ PLANT NUTRITION.

Laws of thermodynamics. The laws of thermodynamics apply to all known phenomena.

The first law of thermodynamics
This law is a statement of the conservation of ENERGY, i.e. energy can neither be created nor destroyed. Energy has many forms – electrical, chemical (the combustion of coal is a release of chemical energy), thermal, nuclear, etc. The first law tells us that it is the total of energy in all its various forms which is a constant. Physical processes only change the distribution of energy, never the sum. However, the first law only tells us energy is conserved, it does not tell us the direction in which processes proceed. This is the province of the second law of thermodynamics.

The second law of thermodynamics
This law specifies the direction in which physical processes proceed. One form of this is that heat transfer takes place spontaneously from a hot body to a cooler body. Note that the opposite process, spontaneous heat flow from a cold body to a hot body (the hot body becoming warmer the cold body becoming colder) *does not violate the first law*. It does not occur because it violates the second law.

The second law is also frequently formulated in terms of order and disorder. It tells us, for instance, that concentrations will disappear, e.g. a sugar lump will dissolve in water, or in general that order becomes disorder. The general statement of the second law is that everything proceeds to a state of maximum disorder, and by implication is less useful in the event of the disorder occurring, e.g. a sugar lump is more useful when it is not dissolved in the water. A kettle of boiling water is more useful than the same amount of water mixed in a bath of cool water. All these examples are manifestations of the second law. It applies to all biological and technological processes and cannot be circumvented. It explains that the final outcome of energy consumption is the eventual production of heat at very low and near useless temperatures, such as the waste heat discharge from power stations or heat losses from the body.

The laws of thermodynamics are often used to determine the ideal efficiency for a given process and explain why the conversion of heat to work cannot be done on a one-for-one basis, i.e. why 1 kilowatt hour of electrical energy output from a generator requires 3 or 4 kilowatt hours of thermal energy input. (⟡ CARNOT EFFICIENCY.)

Leaching. The removal by water of any soluble constituent from the soil or from waste tipped on land. Leaching of domestic refuse tips can give rise to a 'leachate' with a high BIOCHEMICAL OXYGEN DEMAND which is offensive and can spoil amenity of neighbouring watercourses if not ducted away in sealed pipes for treatment. Leaching occurs with soil constituents ranging from nitrate fertilizers (⟡ MODERN FARMING METHODS) to calcium ions used in pH control (⟡ pH).

Leaching of calcium ions, both from the soil and from the leaves of plants which have already absorbed the calcium ions through their roots, is accelerated in areas where there is acidic rainfall due to industrial pollution. When this occurs the plant suffers as calcium is an essential material for cell structure and an enzymatic activator. (⟡ ACID RAIN; ENZYMES.)

Lead (Pb). A HEAVY METAL. A soft, blue-grey, easily worked metal, lead has a multitude of uses, from lead acid accumulators to early water pipes, printer's metal, paint manufacture, and glazes. It is manufactured by roasting the ore (galena) in a furnace. World production is over 3·5 million tonnes annually.

As a pollutant, lead is a systemic agent affecting the brain. It is thought to be associated with mental retardation in infants living in soft-water areas where old lead piping is still in use. Studies of this are being carried out in Glasgow. The soft water carries the lead in solution. Young children are also thought to be at risk of mental retardation from rising lead levels as a result of anti-knock additives in petrol. In addition, lead which is stored in the bones replacing calcium can be remobilized during periods of illness, cortisone therapy, and in old age. Research indicates that lead may blunt the body's defence mechanisms, i.e. the immune system. If this is so, then both the rate of infections by bacteria and viruses and the incidence of cancer can be enhanced by severe exposure (⟡ SYNERGISM).

Blood lead levels for clinical symptoms are set at 0·7 parts per million. However, there is a strong body of opinion that the threshold for children should be 0·3 parts per million, and for adults, 0·4. A normal blood lead level is taken as 0·2 parts per million, but even at this level people can still exhibit lead-induced symptoms.

Lead intake can come from drinking water (World Health Organization limit: 0·05 parts per million), from food, and from inhaled particles from motor exhausts. The particles are less than 1 micron in size, and therefore can enter the lungs easily. A person breathing traffic-congested air may absorb close to the toxic level, excluding any intake from food and drink. Other sources are lead-based paints, and for some children, PICA results in them swallowing paint flakes or chewing toothpaste tubes. (◊CHELATING AGENTS.)

For a detailed account of the effects of lead on mental development, and general health, see D. Bryce Smith, 'Lead pollution – a growing hazard to public health', *Chemistry in Britain*, vol. 7, no. 2, February 1971, pp. 54–6.
For the effects of lead on the body's defence mechanisms, see 'Monitor', *New Scientist*, 18 July 1974.

Lead trap.   The control of lead emissions from motor exhausts can be achieved either by not putting the lead in the petrol in the first place, which could have adverse effects on fuel consumption and motor vehicle performance, or by using a lead trap. This is a device which can reduce lead particulate emissions by about 40 per cent under most vehicle running conditions. The life of the trap materials approaches or betters that of the conventional exhaust system, i.e. 38 000 kilometres (24 000 miles), and thus both can be changed together.

Leaf protein extraction.   This process uses any green leaf as feedstock and yields both a protein-lean but fibre-rich cake and a protein-rich liquor which can be dried for animal feed or used directly. The cake can be fed to ruminants and the liquor to, say, swine. This technique maximizes the use of protein in forage crops as ruminants cannot make best use of all of it as it can be too rich for them.

Lethal concentration.   The concentration of a substance in air or water that can cause death. This is usually expressed as the concentration that is sufficient to kill off 50 per cent of a sample within a certain time, and is abbreviated to LC_{50}.

The toxicity of materials is commonly investigated by the use of survival curves which plot the numbers of population of, say, fish surviving at various times with the concentration of the suspected toxic agent. For a short time almost all fish survive even at high poison concentrations. Very few survive for long periods, hence the LC_{50} concept is very useful as it spans the two extremes. Figure 36 shows survival curves for various concentrations of a toxic material. These curves are plotted as percentage survival against time. The time taken

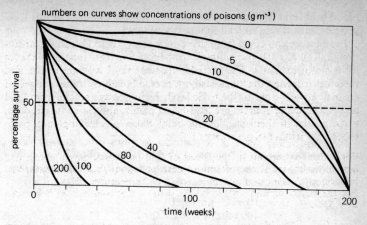

numbers on curves show concentrations of poisons (g m⁻³)

Figure 36. Percentage survival against time for toxicity tests on various concentrations. The points of intersection of the broken line with the curves are lethal concentrations for 50 per cent of the population (LC_{50}).

to kill half the fish is the intersection of the broken line and the concentration curves, e.g. a 75-week LC_{50} would be 20 grammes per cubic metre.

Lethal dose. ◊DOSE.

Lichens. Dual organisms formed by the symbiotic association of two plants, a fungus and an alga (◊SYMBIOSIS). They occur on a variety of surfaces, e.g. tree trunks, rocks, walls and the ground. Nutrition is derived from the dissolved solids in rain which are absorbed through the whole body surface. For this reason lichens are intolerant of poisonous substances and are therefore sensitive to air pollution, especially to sulphur oxides. Classification of surviving lichen flora has been used to monitor maximum air-pollution levels reached over the country.

Lithosphere. The earth's crust, that is, the layers of soil and rock which comprise the earth's crust. When the complete earth's crust is meant, it is often referred to as the hydro-lithosphere. (◊HYDROSPHERE.)

Lignin. The organic glue that holds the cell walls and cellulose fibres of plants together. It is a phenolic polymer which is resistant to biological attack. It forms 25–40 per cent of the wood of trees and must be dissolved chemically in PULP manufacture to obtain high grade pulp.

153

Its presence in agricultural wastes makes many of them unsuitable for feedstocks unless treated. ($\Diamond$ STRAW.)

Logarithms. When two quantities, x and y, are related by an EXPO-NENTIAL CURVE, that is, by the equation

$$y = e^x$$

x may be seen to be the power to which e must be raised to give y. Alternatively x is called the *natural logarithm* of y, and is represented by the symbol 'ln y' or 'log$_e$ y'.

There are many types of logarithms to bases other than e. The logarithms that are most frequently used are those to base 10, in other words, those which result from an equation, $y = 10^x$. These are called *common logarithms* and are represented by the symbol 'log y'. Thus:

$$10 = 10^1, \text{ i.e. } \log 10 = 1$$
$$100 = 10^2, \text{ i.e. } \log 100 = 2$$
$$1000 = 10^3, \text{ i.e. } \log 1000 = 3$$
$$10000 = 10^4, \text{ i.e. } \log 10000 = 4$$

Logarithms to base e and base 10 are chosen to suit the system under consideration. For example, the DECIBEL scale for SOUND measurement uses logarithms to base 10 as does pH ($\Diamond$ pH).

M

Malathion. ◇ORGANOPHOSPHATES.

Marsh gas. ◇METHANE.

Materials, Energy consumption. The energy required in the manufacture of some common materials and manufactured products are shown in the table below. The high energy input and its effects on costs are very apparent and it is obvious why 'alternative technology' is based on both low cost and low energy input materials.

Materials resources. Materials resources, e.g. copper, mercury, tin, lead, differ from energy resources in that, once used, they can be recovered at a cost and used again. Nevertheless, materials are in finite supply and in some cases, like copper, are projected to last in primary form only 40 to 50 years.

On the assumption that no vast new reserves will be discovered and that the restrictions on the availability of ENERGY will not allow very low-grade ores to be worked, the life-time of estimated recoverable reserves (1965 data) are as shown in Figure 37. The life-span of these materials gives little encouragement, and the implications are clear. The consumption of materials should be stabilized if not reduced, and recycling by scrap recovery should be pushed to the highest limits.

Using copper as an example, the world's locatable reserves are about 270 million tonnes, of which 40 per cent is reclaimed scrap. On these figures, copper reserves should last roughly 45 years. If the recycling factor could be increased to 90 per cent, the lifetime could be extended to 270 years (assuming stabilized consumption).

Another possible means of conserving resources is to use alternative materials. For example, aluminium is much more plentiful than copper. However, the energy required to process aluminium is many times that required to process copper. The energy requirement for recycling is much less, however. For steel, the energy consumption for production from scrap is 25 per cent of that required for virgin ores. For aluminium the saving is a staggering 95 per cent, due to the high electricity consumption in bauxite processing.

Energy consumption in basic materials processing*

Material	Energy for unit production (kWh_t tonne^{-1})	Machinery depreciation (kWh_t tonne^{-1})	Transportation (kWh_t tonne^{-1})	Total (kWh_t tonne^{-1})
Steel (rolled)	11 700	700	200	12 600
Aluminium (rolled)	66 000	1000	200	67 200
Copper (rolled or hard drawn)	20 000	800	200	21 000
Silicon, metal and high-grade steel alloys	58 000	1000	200	59 200
Zinc	13 800	700	200	14 700
Lead	12 000	700	200	12 900
Miscellaneous electrically produced metals	50 000	1000	200	51 200
Titanium (rolled)	140 000	1000	200	141 200
Cement	2000	50	50	2300
Sand and gravel	18	1	2 (short distance hauling)	21
Inorganic chemicals	2400	100	200	2700
Glass (plate finished)	6700	300	200	7200
Plastics	2400	300	200	2900
Paper	5900	300	200	6400
Lumber	1·47 per board foot	0·02 per board foot	0·02 per board foot	1·51 per board foot
Coal	40	2	—	42

*Accuracy ± 20 per cent.
From 'Economical use of energy and materials', *Environment*, vol. 14, no. 5, June 1972, p. 14.

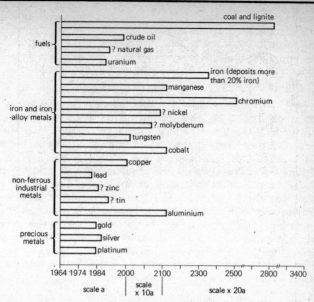

Figure 37. Estimated world mineral reserves based on current technology and consumption rates. Future discoveries have not been taken into account. (From P. Cloud, 'Realities of mineral distribution', *Texas Quarterly*, vol. 11, 1968, pp. 103–26.)

The USA with 6 per cent of the world's population consumes up to 30 per cent of the world's material resources. If the underdeveloped countries wished to have this level of per capita production, then the world's production of iron, copper and lead would have to rise six- to eight-fold. The resources are simply not available to give the underdeveloped countries this level of production.

See P. Chapman, 'Energy costs of producing copper and aluminium from primary sources', *New Scientist*, vol. 8, no. 2, 12 December 1974, p. 107.

Magnetohydrodynamic generator (MHD). A device for DIRECT ENERGY CONVERSION which generates electricity by the passage of a gas at high temperature and pressure through a magnetic field. The gas is ionized and this is enhanced by seeding with caesium or potassium. The gas is then passed through a nozzle at right angles to a magnetic field. The magnetohydrodynamic generation is essentially a 'topping' cycle – that is, a means of extracting greater electrical energy from a given amount of fuel by burning the fuel at as high a temperature as possible (2500°C or greater) followed by a conventional steam-turbine generation at a lower temperature.

Magnetohydrodynamic generation received great attention in the 1950s and 1960s but has now dropped from favour on grounds of capital cost and scale, as it can only be usefully employed in the range from 10 megawatts upwards.

Malting. This is, essentially, the speeded-up process of barley seed germination. The barley is soaked (for up to two days), then spread in layers (up to 1 ft) in a warm temperature to encourage growth which is allowed to proceed for four to six days. Germination is halted by kilning and the end product is malt which may be used for brewing beer.

During the malting process the cellulose walls around the starch cells are broken down by hydrolytic ENZYMES. Kilning stops this process but the temperature and time are regulated so that only minimum enzyme destruction occurs. During the first process of brewing (mashing) the hydrolytic enzymes (mainly diastase) recommence work to break down the starch in the malt to water-soluble products including maltose, sucrose and glucose. Brewers' yeasts are able to convert maltose and sucrose to glucose which is then fermented to produce alcohol. (◊ FERMENTATION.)

Maximum allowable concentration (MAC). In the USA, the upper limit for the concentration of noxious or toxic emissions in workplaces. Exposure to the MAC of a pollutant should not cause any distress to anyone except the odd sensitive individual. The usual guidelines adopted in the UK are the THRESHOLD LIMITING VALUES (TLV). (◊ THREE MINUTE MEAN CONCENTRATION.)

Maximum sustainable yield. The maximum yield that can be obtained from a given crop or species if it is to maintain equilibrium. The management of a forest or fishery or farm should avoid overgrazing, overcropping or overharvesting. Thus, fishery catches should be strictly controlled so that the fish population can have a sufficient breeding mass and thus give a sustained yield for future generations. This philosophy is particularly applicable to whaling (◊ WHALE HARVESTS), herring and salmon fishing, but is relevant to many areas, such as forestry, and is basically the application of good stewardship of natural resources that cannot regenerate if exploited too far.

The exploitation of the herring is a particularly apt example. The last stock of herring left in European waters is off the west coast of Scotland. In the season over 300 ships with nets as large as St Paul's Cathedral have fished there. The fish are 'hoovered' by suction techniques into the holds and pulped for fish meal. Already the Icelandic and Norwegian fisheries have been killed off and in 1975 Norwegian boats hoovered up to 12 miles off the shores of the Shetlands. Unfortunately these countries have no conservation policy; they are geared to industrial fishing. Norway alone has 500 industrial ships catching fish for 250 fishmeal factories.

The destruction of the North Sea herring is now recognized as one of the most short-sighted man-induced examples of ecological upsets this century. Yields have dropped 300 per cent in the last 20 years.

Control of net sizes and fishing techniques is required so that spawning can take place and young fish left alone. Such action may in time allow breeding stock to recover. If these controls are not implemented soon, the North Sea herring will go the way of the North American buffalo.

Counterparts abound everywhere: a river should not be so depleted of DISSOLVED OXYGEN by domestic or industrial pollution that it cannot recover. The land should not be 'mined' for short-term agricultural profit. The COMMONS must be husbanded and good husbandry enforced.

Mean annual temperature (global).   This is the mean temperature of the earth (averaged over a year) for individual latitudes leading to a composite value which is a measure of the earth's radiation balance. If more radiation is absorbed by the land mass and oceans this year than was absorbed last year without a compensating change in the outgoing radiation, the mean annual temperature will rise a fraction of a degree. If the opposite process occurs, it will drop.

The mean annual temperature in both hemispheres has changed in the last 100 years. In the northern hemisphere between 1880 and 1940 it rose 0·6°C, and since 1940 it has dropped by 0·3°C. A corresponding but smaller fluctuation has been observed in the southern hemisphere.

The earth's ALBEDO is a significant factor in the mean annual temperature fluctuations, as is the degree of airborne dusts and fumes from industrial and quarrying operations.

Means, Best practicable.   According to the Alkali Act 1906, a scheduled process must be provided with the *best practicable means* for preventing the escape of noxious or offensive gases and smoke, grit and dust to the atmosphere and for rendering such gases, where necessarily discharged, harmless and inoffensive. The words 'best practicable means' are often used to describe the whole approach of British anti-pollution legislation towards industrial emission: the cost of pollution abatement and its effect on the viability of industry are taken into account. Thus, while better pollution control can often be achieved, the 'best practicable means' philosophy can allow lower standards of control to keep a sector of industry viable, having regard to established practice, the area in which it is sited, etc. In effect this may mean that no firm is legally compelled to modify its activities or install pollution-abatement equipment until it is economically convenient to the firm to do so. In the meantime, the social costs of the consequent pollution continue to be borne by the community. The firm may well be urged to investigate means of suppressing pollution, but may not be forced to do so. (◇EXPOSURE–DOSE EFFECT RELATIONSHIP; THRESHOLD LIMITING VALUE.)

Mercaptans. A family of foul-smelling sulphur compounds produced by decaying organic matter, emitted at sewage works, food-processing plants, brick-making works and oil refineries. The most offensive, ethyl mercaptan (C_2H_5SH; boiling point 30°C) can be smelt at concentrations as low as 1 part in 50000 million in air.

Mercury (Hg). A HEAVY METAL that exists as a liquid at normal temperatures; atomic weight 200·59. It is extracted from an ore (cinnabar). Mercury is used as a catalyst in many industrial processes as well as in barometers, thermometers, etc. Over 10000 tonnes are produced annually.

Provisional tolerable weekly intake for man

Substance	(mg per person)	(mg kg^{-1} body-weight)
Mercury		
Total mercury	0·3	0·005
Methyl mercury		
(expressed as mercury)	0·2	0·0033
Lead[1]	3	0·05
Cadmium	0·4–0·5	0·0067–0·0083

1. These intake levels do not apply to infants and children.

From Food and Agriculture Organization, *Evaluation of Mercury, Lead, Cadmium and the Food Additives Amaranth, Diethylprocarbonate and Octy Gallate*, Rome, 1973.

As a pollutant it is a systemic agent, affecting the brain, kidneys and bowels (◇MINAMATA DISEASE). The organic forms, e.g. methyl mercury, are particularly toxic. World Health Organization food regulations state 0·05 part per million as the highest allowable concentration of mercury in foodstuffs.

Inshore fish in the Mersey estuary and Morecambe Bay contain up to 1·1 mg kg^{-1}. Fish from distant waters contain an average 0·06 mg kg^{-1} (Ministry of Agriculture, Fisheries and Food, *Survey of Mercury in*

Food, H M S O, 1971). It is possible that people with very restricted diets of local fish (◊CRITICAL GROUP) could ingest more than the provisional tolerable weekly intake for man as shown in the following table for mercury, lead and cadmium.

The term 'tolerable' signifies permissibility rather than acceptability since the intake of a contaminant is unavoidably associated with the consumption of otherwise wholesome and nutritious foods: the term 'provisional' expresses the tentative nature of the evaluation (FAO, 1973, p. 6).

The provisional tolerable weekly intake is used for metallic contaminants in food as an *acceptable daily intake* cannot be readily determined.

There is a paucity of information on the effects of sub-lethal doses of heavy metals but, in view of the now-suspected properties of LEAD, there is obviously a strong case for strict control of all heavy-metal discharges and emissions.

Emission sources are coal burning (where mercury is always present), the paper industry and chemical plants. (◊CHLOR-ALKALI PROCESS.)

Methane (Marsh gas; CH$_4$). A gas which can be easily liquefied. It is chiefly used as a fuel. Methane occurs naturally in oil wells and as a result of bacteriological decomposition, e.g. the anaerobic digestion of sewage SLUDGE. It can also be synthesized.

There is a major environmental school which believes that methane is the fuel of the future and that its production should be encouraged by every household using a simple anaerobic digester. Methane produced by bacterial action on faeces and household organic refuse will provide 10–25 per cent of a household's energy requirements. To ensure rapid bacterial action or FERMENTATION, the anaerobic treatment should be conducted at a temperature of 35°C which in temperate climates means some form of heating is required. The equipment is simple and standard references such as that of the Doubleday Research Institute will enable a plant to be built.

Present work has shown that methane is a suitable carbon source for bacterial growth and SINGLE CELL PROTEIN manufacture. It is also the raw feedstock for methanol production which may be used as a fermentation substrate for single cell protein production.

Methane Fuel for the Future, Henry Doubleday Research Association, Bocking, Braintree, Essex.

Methanol (Methyl alcohol or wood alcohol; CH$_3$OH). A chemical, 'methylated spirits', that can be produced by destructive distillation of coal or wood. Methanol is usually synthesized from carbon monoxide and hydrogen or from METHANE. It is used as an intermediary by the chemical industry, and also as a solvent and denaturant.

One important and growing use is as a substrate for the production of protein; a class of bacteria has been isolated that can utilize it. It is also being actively promulgated as a motor fuel to conserve oil reserves, either on its own or mixed with petrol in proportions between 15–30: 85–70 per cent. It is claimed that higher mileages result, and that 50 per cent less carbon monoxide is produced from such fuel mixtures. In the USA a strong lobby exists for its manufacture from domestic refuse, thereby conserving oil resources and achieving waste recycling.

Methaemoglobinaemia. A disorder, known as the 'blue baby disease', which affects the oxygen-carrying capacity of the blood. It is associated with drinking water containing nitrogen in the form of nitrates. The amount of nitrate per litre of water at intakes on the Thames and the Lee has, since 1972, approached (and for the Lee exceeded) the 50 milligramme per litre nitrate (or $11 \cdot 3$ mg l^{-1} as nitrogen) limit considered to be safe. Above this limit, bottle-fed babies may develop methaemoglobinaemia, while above a concentration of 20 milligrammes per litre as nitrogen, the adult population may be at risk. So far, London's water supply has been kept well below the safe limit by diluting the Thames and Lee water with water from reservoirs.

Modern farming techniques involving heavy applications of nitrogen FERTILIZER, extensive cultivation of leguminous plants such as peas and beans, and sewage works effluent are all contributing to the high nitrate levels. The residues from agriculture are present in the ground in large quantities, and modern land drainage techniques mean that they will eventually find their way into the water supply. (◇MODERN FARMING METHODS; SOIL, FERTILITY AND EROSION OF.)

For further information on the presence of nitrate in the Thames and the Lee, see R. Scorer, 'Nitrogen: a problem of increasing dilution', *New Scientist*, 25 April 1974.

Microbes. Microscopic organisms. The term usually refers to bacteria. Many are pathogenic, e.g. salmonella, which is associated with food poisoning in man.

Micron. One-millionth of a metre. It is commonly used for particle sizing. Symbol μm in SI units.

Minamata disease. Minamata is a town on the west coast of Kyushu island (Japan) where an extreme case of HEAVY METAL poisoning from methyl MERCURY ingested in the staple fish diet of the inhabitants caused severe disablement and death: 43 deaths and 68 major disablements were recorded between 1953 and 1956.

The symptoms include numbness in fingers and lips and difficulty in speech and hearing. There is a marked inability to control limbs, followed by seizures. Children and old people are particularly vulnerable, as is the case with all heavy metals ingestion.

The source of mercury in the bay was eventually traced to a PVC plant which used mercuric sulphate (an inorganic chemical) as a CATALYST. This was subsequently converted by marine life to the methyl form. Minamata is now a classic case of industrial pollution and subsequent evasion of responsibility.

Minerals. ▷NUTRITIONAL REQUIREMENTS, HUMAN.

Mists. Microscopic liquid droplets suspended in air. Their diameter is less than 2 microns. (▷FOG.)

Modern farming methods. Good farming has a minimum effect on the environment by retaining as much natural diversity as possible – an effect which can be achieved by planting a variety of crops in rotation in small fields enclosed by hedges. Such practice is recognized as sound by both farmers and ecologists, yet the demands of a maximum guaranteed return on capital as the sole criterion of 'efficiency' is producing obvious and grave effects such as:

The wholesale removal of hedges.

The replacement of manpower by machines and chemicals.

The conversion of large areas of countryside to specialized production (▷MONOCULTURE).

Widespread pollution of watercourses by 'strong' (high BIOCHEMICAL OXYGEN DEMAND) discharges from dairy farms, especially in the West Country.

The removal of livestock from the fields to factory-type units, where they are fed vegetable PROTEIN and fishmeal imported from underdeveloped countries.

The replacement of animal waste products as a food for the soil by artificial FERTILIZERS. The long-term application of nitrogenous fertilizers can eventually lead to the appearance of nitrates in aquifers used for water supplies.

The compaction of soil by heavy farm machinery, enhancing run-off of the fertilizer so that it becomes an additional pollutant of rivers (▷EUTROPHICATION).

The widespread use of ANTIBIOTICS in an attempt to counteract diseases that can proliferate among large numbers of animals kept in close proximity and the use of hormones for growth promotion.

The appearance of bacterial strains that are immune to antibiotics. It has been found that bacterial resistance to antibiotics is induced in bacteria harmless to people, but there is a mechanism for this resistance to be transferred to other bacteria which are dangerous to people. The Swann Committee has recommended a ban on most of the antibiotics used in this way. So far, no such ban has been imposed. Thus the 'Agricultural Revolution'. (◊AGRICULTURAL ECONOMICS; AGRICULTURE, ENERGY AND EFFICIENCY ASPECTS.)

Swann Report: *Report of the Joint Committee on the Use of Antibiotics in Animal Husbandry and Veterinary Medicine*, HMSO, 1969.

Molecular formula.   ◊ELEMENT.

Mono-crops.   The culture of a single plant type, e.g. cereal crops. Mono-crops cannot survive without man's intervention, his cultivation, and weed- and pest-killers. Ecosystems are simplified and vast areas have no stable community because of such practices. (◊ECOSYSTEMS.)

Mono-culture.   The planting of single crop species exclusively. Mono-culture encourages increases in the population of plant-eating insects and in diseases, and discourages the presence of their natural predators. For example, cereal production has been followed exclusively on the same fields in the East Midlands for 20 to 30 years. The build-up of disease is a constant threat, as the normal rotation of crops, which eliminates the build-up of soil-borne diseases, is not practised under this system. (◊ECOSYSTEM.)

Monomer.   A chemical compound consisting of single molecules, as opposed to a POLYMER, the molecules of which are built up by the repeated joining of monomer molecules.

Mosses.   Multicellular, CHLOROPHYLL-bearing plants. Mosses are not complex but have a highly developed reproductive system. They occur in damp conditions, such as in woods and also on walls. The stems contain a central core of elongated cells which act as ion-exchange membranes. The intricate branching system with many leaves provides large surface areas on which particles can be trapped. This structure means that mosses can be used to concentrate airborne pollutants. Thus the presence (or absence) of certain species of mosses (and LICHENS) indicates the absence (or presence) of airborne pollutants such as sulphur dioxide, fluorides, and copper, lead and zinc fumes, etc. Where the mosses are absent, mossbags – i.e. bags of dry dead moss suspended in nylon nets – can be used. As the accumulation rate of the metals is a function of air concentration of the pollutant(s), the average concentration over the period of exposure can be found by analysis of the mossbags.

Multiple flues.   In large power stations or process plants, where there are separate boilers, the best practice for good plume dilution and dispersion is to take all gas discharge streams to one chimney with multiple flues so that emission velocity (◊ PLUME RISE) is maintained over a wide working range and interference from a multitude of chimneys, each with its own small plume, is avoided. The use of multiple flues means that one flue discharge does not affect the combustion draught conditions of an adjacent flue.

Mutagen.   Something (e.g. a chemical or radiation) that changes the genetic material (chromosomes) that is transferred to the daughter cell when cell division occurs. The result is that the new cells have changed characteristics.

The mutagen can act on a *germ* cell, e.g. sperm of man or that of any other sexually reproducing organism (rats, mice, fruit flies), and some of the offspring will carry the mutant genes in all their cells. A mutagen may also affect *somatic* cells, in which case the effects are to the person concerned and not to offspring. These effects depend on the type of cell affected. One effect could be to make cells grow and multiply faster than they can be removed by the blood; if the white cells are affected, the result is leukaemia. Another effect could be to start up cell divisions in cells that do not normally divide; if the division products displace or invade normal tissues, the result is a cancer. Both types of mutation are discussed further; ◊ IONIZING RADIATION.

Mutagens may also be carcinogenic or teratogenic, and therefore any substance that is found to be mutagenic must be tested for its carcinogenic and teratogenic properties as well (◊ CARCINOGEN; TERATO-GEN). Although mutagens may not be carcinogens or teratogens, they must be considered suspect until cleared. *This conservative viewpoint is essential if human health is to be protected.* There is often a major time-lag between contact with a mutagenic agent and the onset of cancer in man. For example, VINYL CHLORIDE was only found to be carcinogenic after long use. It is now becoming apparent that many substances in common use are suspected as being carcinogenic, e.g. hair dyes. Because of the latency period before cancer manifests itself, these suspected substances may continue in use for many years. The principal area for research activity and concern is the use of the growing class of chemicals that affect nucleic acids, the basic chromosome component.

The World Health Organization Scientific Group on the Evaluation and Testing of Drugs for Mutagenicity stated that the following should receive special attention:

(1) compounds that are chemically, pharmacologically and biochemically related to known or suspected mutagens;

(2) compounds that exhibit certain toxic effects in animals, such as depression of bone marrow; inhibiting of spermatogenisis or oogenesis; inhibition of mitosis; teratogenic effects, carcinogenic effects, causation of sterility or semi-sterility in reproduction studies; stimulation or inhibition of growth or synthetic activity of a specific cell or organ; inhibition of immune response; and

(3) compounds that are likely to be continuously absorbed into the body and retained by it for long periods.

The testing should be done with mammalians, e.g. rodents, so that an indication of the potential effects on man may be determined.

A mutagenic index is derived from the tests which reflects the percentage of dominant *lethal* mutations in the experimental group.

The finding of mutations in one species does not mean that it is mutagenic in all species, but a positive finding should certainly be considered as an indication of potential mutagenic activity in man.

Mutagenic index. ◊MUTAGENS.

Mutant genes. ◊IONIZING RADIATION.

Mutation. The random alterations of reproductive cells of organisms which result in changes in the other cells of the organism. If a particular mutation protects the organism in a hostile environment, then the mutant has a good chance of survival. (◊GENETIC EROSION.)

N

Natick process. A cellulose decomposition process which uses an enzyme produced by a fungus, TRICHODERMA VIRIDE, for glucose and single cell protein production. It was developed by the US Army Natick Laboratories. For a description of the process, ◊ENZYME TECHNOLOGY.

Natural background radiation. ◊IONIZING RADIATION.

Natural turnover. The annual throughput of material 'processed' by a natural system in equilibrium (◊CARBON CYCLE). The natural turnover can be used to scale man-made inputs of the same material in order to assess any risks that may occur if the man-made input swamps the natural system and destroys its stability.

Nephritis. Inflammation of the kidneys which can be caused by drinking water with lead in solution; e.g. rainwater collected from lead-painted roofs.

Nitrate preservatives, Association with cancer. Recent research carried out at Imperial College, London, and supported by the Cancer Research Campaign, indicates that coffee, combined with certain foods such as bacon, cheese and corned beef, may be capable of producing cancer. Coffee contains chlorogenic acid, which reacts with the nitrate preservatives put into pork, cheese and corned beef to produce nitrosamines, an established cancer agent.

New lower permissible levels for nitrate additives in food are soon to be issued by the American government – levels which will be below what is at present allowed in Britain. Under the new standards, the ceiling for nitrate curing agents applied to most meats will be 156 parts per million. Because some nitrates are lost during processing, lower levels – between 50 and 125 parts per million – will be set for meat leaving the plant. The current ceiling at this stage is 200 parts per million, which is the limit in Britain.

In Britain, most manufacturers are steadily decreasing the amount of such additives, but since the absence of preservatives reduces a product's shelf-life and increases the chances of botulism, there is obviously a minimum concentration below which it would be unwise to go.

Nitric oxide. ⬦NITROGEN OXIDES.

Nitrification. Conversion of nitrogenous matter into nitrates by bacteria, especially in soil. (⬦NITROGEN CYCLE.)

Nitrogen cycle. The atmosphere contains 78 per cent by volume of nitrogen (N_2) and 21 per cent by volume of oxygen (O_2), yet nitrogen is not a common element on earth. It is an essential component for plant growth and proteins, yet it is chemically very inactive and before it can be incorporated by the vast majority of the biomass, it must be fixed. The fixation of nitrogen is its incorporation in a combined form such as ammonia (NH_3) or nitrate (NO_3), whereby it can be used by plants or animals. This fixation can be carried out industrially or naturally by the action of bacteria.

As the industrial fixation of nitrogen now matches the bacterial fixation rate, this is an example of man matching a natural cycle, so both routes must be discussed.

Industrial fixation (Haber process)
The fixation of nitrogen industrially is carried out at 500°C and high pressure (200 atmospheres) and involves (a) the reforming of methane to provide hydrogen, (b) the introduction of atmospheric nitrogen and oxygen. The oxygen reacts to form carbon monoxide and thereafter a catalytic reaction combines nitrogen and hydrogen to form ammonia, which can then be readily converted to nitric acid, and then ammonium nitrate, a fertilizer of wide applicability. Current nitrogen fixation rates are around 40 million tonnes per year and are increasing.

Biological fixation
Nature accomplishes nitrogen fixation by means of nitrogen-fixing bacteria. The components of the natural nitrogen cycle are shown in Figure 38. They form a very intricate chain of interlocking activities and are essential to the maintenance of the atmospheric composition.

We start with plants such as peas, beans and clover (legume family) which have nitrogen-fixing bacteria living symbiotically in nodules on their roots and can fix as much as 0·24 to 0·36 tonnes nitrogen per hectare. The plant supplies the bacteria with food and energy, and the bacteria fix atmospheric nitrogen in the form of soluble compounds, some of which are excreted into the soil while the rest supply the plant with essential nitrates. The fixed nitrogen is incorporated in plant protein and if eaten becomes reincorporated as new proteins in the animal.

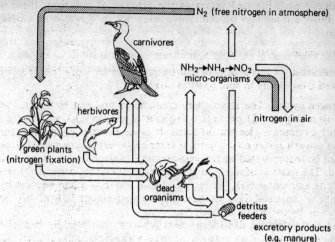

N₂ (free nitrogen in atmosphere)

carnivores

$NH_2 \rightarrow NH_4 \rightarrow NO_2$
micro-organisms

nitrogen in air

herbivores

green plants
(nitrogen fixation)

dead
organisms

detritus
feeders

excretory products
(e.g. manure)

Figure 38. The nitrogen cycle.

Eventually this protein returns to the soil when the animal or plant dies (there is also a contribution from defecation) and is decomposed by bacterial action into its component amino acids. In aerobic soil conditions many bacteria oxidize these amino acids to carbon dioxide, water or ammonia. Two additional micro-organisms – the nitrifying bacteria – convert the ammonia to nitrite (NO_2) and thence to nitrate (NO_3). Once in the nitrate form it is again available for plant food. Thus, the cycle nitrate to protein, protein to ammonia, ammonia to nitrite, nitrite to nitrate can be repeated. Similarly animal excreta are decomposed to ammonia and converted to nitrate by the nitrifying bacteria.

Superimposed on the cycle are the denitrifying bacteria which decompose (at a smaller rate than fixation) the nitrites or nitrates to molecular nitrogen (N_2) to the atmosphere. Without these denitrifying bacteria, most of the nitrogen would be locked up as run-off in the oceans or in sediments.

As man fixes 40 million tonnes of nitrogen per year with estimates showing that biological fixation is of the same magnitude, and as the denitrifying bacterial activities were in balance with the biological fixation, it must be assumed that much of the man-made nitrates eventually end up as run-off in lakes, rivers or seas where EUTROPHICA-TION can result. The benefits of nitrate application are obvious in increased crop yields; however, the ecological implication of interfering with natural cycles have not been explored in sufficient depth. Industrial nitrogen fixation is expected to total 100 million tonnes per year by the year 2000 to cope with ever-increasing demands for food from finite land resources. In fact, man can defeat nature in some cases, as the application of nitrogenous fertilizers stops the action of the nitrogen-fixing bacteria.

Breeding developments are in hand to induce these bacteria to live in symbiotic relation with other food crops such as cereals.

The nitrogen cycle is very closely linked to those of carbon and oxygen, as some of the bacteria involved oxidize organic matter in the soil. There is, thus, a simultaneous cycling of carbon, oxygen and nitrogen.

In aquatic environments, nitrogen fixation can be accomplished by blue-green algae which can make some waters highly productive regions, rich in fish life because of the availability of food.

Nitrogen dioxide. ▷NITROGEN OXIDES.

Nitrogen-fixing bacteria. ▷NITROGEN CYCLE.

Nitrogen oxides. These are collectively referred to as NO_x. Three are of importance:

Nitrous oxide (N_2O) is a colourless gas used as an anaesthetic; background concentration is 0.25 part per million. It is mainly formed by soil bacteria in decomposing nitrogenous material.

Nitric oxide (NO) is colourless, formed during the high temperature combustion of fuels which allow the nitrogen in the air to combine directly with oxygen. This takes place at temperatures above $1600°C$, but if the fuel also contains nitrogen, nitric oxide will be produced at temperatures above $1300°C$.

When cooled rapidly it can form nitrogen dioxide (NO_2), which is a reddish-brown, highly toxic gas with a pungent odour. It is one of the seven known nitrogen oxides which participate in PHOTOCHEMICAL SMOGS and primarily affect the respiratory system. Nitrogen dioxide is extremely poisonous and can pass into the lungs and form nitrous acid (HNO_2) and nitric acid (HNO_3), both of which attack the mucous lining. Nitrogen dioxide is also thought to act as a plant-growth retardant at normal atmospheric concentrations.

The rough magnitude of nitrous oxides emissions for Los Angeles, using average US car figures of 15.6 grammes of nitrous oxides per litre of fuel consumed, is over 400 tonnes of nitrous oxides per day. Altogether the emissions of nitrous oxides need careful watching.

Severe legislative measures are under way in the USA for their control. (⌑AUTOMOBILE EMISSIONS.)

Nitrosamines. An established cancer agent. (⌑NITRATE PRESERVA-TIVES.)

Noise. SOUND that is socially or medically undesirable, i.e. any sound that intrudes, disturbs or annoys. Very high levels of sound can cause hearing damage. Noise terminology is complex as the physical properties of the sound and the mental processes by which it is heard and reacted to must be considered. (⌑HEARING; ROAD TRAFFIC NOISE; AIRCRAFT NOISE; INDUSTRIAL NOISE MEASUREMENT; NOISE INDICES.)

Noise and number index (NNI). Index of air traffic noise. For a number of aircraft, N, heard in a day, NNI is given by the equation:

$$NNI = \text{average peak perceived noise level} + 15 \log N - 80,$$

where the average noise level is the logarithmic average.

Daytime NNI is the average NNI value during the summertime (mid-June to mid-September) for a daytime period from 06.00 until 18.00 hours. Note that 80 is subtracted from the NNI as this is the supposed zero annoyance level. (⌑AIRCRAFT NOISE; LOGARITHMS.)

Noise control. There are three identifiable areas for noise control: source, path, and receiver. A source generates the noise. The path is the source–receiver separation and is characterized by the nature of the intervening space. The receiver is the listener. Noise control can be attempted at any one or all three of these areas.

1. *Control at source*. This is the most widely used form of noise control, e.g. motor vehicles are fitted with silencers and are required to emit less than a maximum permitted noise level (emission standard). For heavy vehicles in particular, control at source can mean tackling any or all of the components shown in the table on page 172.

2. *Noise control between source and receiver*. In many instances the source cannot be significantly altered, e.g. traffic noise, and therefore sound insulation such as double glazing, fan ventilation instead of open windows and air bricks, and acoustic tiles must be employed. Noise barriers can also be erected. These should be as close to the source as possible and have a length of at least ten times the shortest distance between source and observer so that the sound waves do not merely bend round the ends of the barrier and so reach the observer. They will bend over the top as well but this cannot be easily avoided.

	Possible control measures	
Engine casing	Split in two to balance and 'break' transmission vibration	
Engine inlet	Smooth passages and enlarge	
Engine exhaust	Redesign silencer	Total
Transmission & drive train	Anti-vibration Redesign gear train	emitted airborne
Fan	Use variable pitch blades	noise
Tyre/roadway	Redesign suspension Improve road surface	
Aerodynamics	Redesign body shell contours	

The effectiveness of sound barriers or acoustic insulation is proportional to their density. The denser the material the greater the reduction of SOUND. The UK recommendation is that 70 dB(A) is the limit of the acceptable, rather than a standard of what is desirable, for external sound levels from traffic noise for domestic areas (◊DECIBELS A-SCALE).

3. *Receiver control*. This means that ear protectors are required and these should be chosen for the particular sound pressure level that is to be reduced to a safe level. Ear protectors are mainly used in an industrial environment and are particularly important if NOISE-INDUCED HEARING LOSS is to be avoided.

Noise indices. Measures of the disturbing qualities of noise – loudness, variations in time, whines, bangs, etc. – and the average subjective response. Individual responses to noise vary; noise indices are average measures. They are used in planning residential developments, motorways, etc. TNI is the traffic noise index for road traffic; L_{10} (18-hour) index is used in ROAD TRAFFIC NOISE legislation; NNI is the NOISE AND NUMBER INDEX for air traffic; CNL is the CORRECTED NOISE LEVEL for noise from industrial premises.

Noise-induced hearing loss. Damage to the ear not caused by ageing or accident or disease. This is an insidious affliction. Many industries are inherently noisy – foundries, boiler shops, shipyards, press rooms. A worker first entering such an environment spends some time getting acclimatized and then may eventually accept the noise levels to which he or she is subjected. This casual acceptance encourages in both workers and management an attitude that noise is simply part of the job. The first effects of exposure to excessive noise are ringing in the ears and a dulling of the hearing – temporary threshold shift, i.e. a 'louder' level of noise is required to conduct a conversation. These effects are often temporary, but if there is continued exposure or infrequent gaps between exposures, the ears do not recover and there is permanent damage. The inner ear can be damaged and in extreme cases the eardrum can be ruptured.

Industrial deafness is usually associated with inner-ear damage which may be unnoticed for many years until it is chronic. The usual result is that frequencies above 4000 hertz such as a high-pitched whistle or the full range of music cannot be detected. As speech is usually in the 500–2000 Hz range, it is only when an affected individual cannot follow speech so readily that the damage is detected. To assess noise-induced hearing loss an audiometer is used which compares the subject's hearing threshold with that of a normal hearing individual. Thresholds are usually measured at 0·25, 0·5, 1, 2, 3, 4, 5, 6 and 8 kilohertz. Thus at, say, 2 kHz (speech frequency), if the subject is asked to identify the intensity at which he just hears 2 kHz and if this is 30 decibels, then the hearing level is said to be 30 dB at 2 kHz. Now normal hearing intensity at 2 kHz is 0 dB, so that the subject's threshold differs by 30 dB. An audiogram can be obtained which plots noise-induced threshold shift against frequency as shown in Figure 39.

The degree of handicap is given in the first table. Ordinary conversation is not the only criterion: the social dimension of hearing loss is

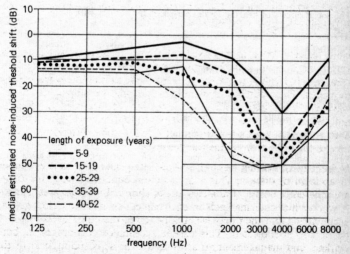

Figure 39. Noise-induced hearing loss as a function of length of exposure for weavers.

extremely important and this is shown in the second table. Clearly excessive noise is also a damaging pollutant.

Classification of handicap due to hearing loss

Class	Degree of handicap	Average hearing level (dB)	Ability to understand ordinary speech
A	Not significant	Less than 25	No significant difficulty with faint speech
B	Slight	25 to less than 40	Difficulty only with faint speech
C	Mild	40 to less than 55	Frequent difficulty with normal speech
D	Marked	55 to less than 70	Frequent difficulty with loud speech
E	Severe	70 to less than 90	Shouted or amplified speech only understood
F	Extreme	90	Usually even amplified speech not understood

From W. Burns, *Noise and Man*, John Murray, 1968; revised edn, 1972.

The social effects of hearing loss experienced by weavers with long exposure to noise

Social effects	Weavers	Control*
Difficulty in understanding family/friends	77%	15%
Difficulty in understanding strangers	80%	10%
Difficulty in use of telephone	64%	5%
Difficulty at public meetings, church, etc.	72%	6%
Own estimate of hearing below normal	81%	5%

* In this table a population of weavers who had been working in the trade for a number of years is compared with a control population who did not have a history of severe noise exposure.

From W. Taylor *et al.*, *Proceedings of a Conference on Occupational Noise in Medicine*, National Physical Laboratory, 1970.

Noise measurement. ◇DECIBEL; SOUND.

Novel protein. Alternative source of protein to meat, derived from soya beans. Other possible sources are yeasts, fungi, bacteria, algae and leaves. (◇SINGLE CELL PROTEIN; LEAF PROTEIN EXTRACTION.)

Nuclear energy. There are potentially two principal processes for obtaining energy from nuclear sources – fission and fusion. (Fusion has not yet been engineered as a continuous process.)

Fission uses the energy released in a controlled nuclear reaction whereby an isotope (uranium–235) is split by the capture of neutrons, energy (plus more neutrons) is released and mass consumed in the process. One gramme of uranium–235 consumed in a fission reaction will release 81 900 million ($8 \cdot 19 \times 10^{10}$) joules or the equivalent of the heat of combustion of $2 \cdot 7$ tonnes of coal or $13 \cdot 7$ barrels of crude oil – hence nuclear power has much to commend it in energy terms. A 1000-megawatt nuclear power station would consume about 3 kilogrammes of ^{235}U per day. While seemingly small, this is a large amount of uranium in relation to the total supplies, and the prospects for nuclear power would be relatively short-lived were it not for the various other types of reaction (such as the breeder reactor) and the possibility of an alternative long-term nuclear energy supply – fusion.

Fusion relies on the energy release when a heavier element is formed by the fusion of lighter ones. The sun's energy is a fusion process resulting from the formation of helium by the fusion of atoms of one or more of the hydrogen isotopes, i.e. hydrogen (H), deuterium (D) and tritium (T). As in fission, there is a loss of mass and it is this mass difference that appears as energy. Thus, the fusion reaction revolves round the hydrogen isotopes and, indeed, the energy released in an uncontrolled explosive manner by the fusion of D and T is the basis of the H-bomb.

The potential of fusion power is illustrated by considering 1 cubic metre of water which ordinarily contains 1 D atom for each 6500 H atoms and has a potential fusion energy (D–D fusion) of 8160 000 million ($8 \cdot 16 \times 10^{12}$) joules or the heat of combustion of 269 tonnes of coal or 1360 barrels of crude oil.

1 cubic kilometre = 1000 million cubic metres, so that 1 cubic kilometre of water contains the fusion (D–D) potential of 269 thousand million tonnes of coal or 1330 thousand million barrels of crude oil, which approximates to the lower estimate of the world reserves of crude oil (⟡ENERGY RESOURCES).

The total volume of the ocean is about 1500 million cubic kilometres, and if 1 per cent of this were used, the potential energy release would be sufficient for 5 000 000 times that of the world's initial coal and oil supplies (see Hubbert, 1969). The reasons for the quest for controllable fusion power are now evident.

For the provision of nuclear energy as currently practised, i.e.

fission, ⟡NUCLEAR REACTOR DESIGNS. (⟡⟡IONIZING RADIATION; NUCLEAR REACTOR WASTES.)

M. K. Hubbert, "Energy resources", *Resources and Man*, W. H. Freeman, 1969, ch. 8.

Nuclear fission. In spite of the intense and very costly development programmes of the last two decades, the much heralded future of power from nuclear fission is still not in sight. Great Britain still generates only one-tenth of her electric power from nuclear fission (the extraction of heat either directly from fissile isotopes or indirectly from fertile isotopes). Nevertheless, the proponents of nuclear power have been predicting growth rates corresponding to doubling times of a few years which would, in practice, be quite impossible to achieve (Chapman, 1974).

The problems of assessing the relative merits and demerits of an extensive nuclear fission technology are of an ethical nature and cannot simply be weighed by cost–benefit analysis. There is a tendency to overemphasize the short-term benefits and undervalue the long-term hazards (see Kneese, 1973). In attempting to assess the risks associated with a fissile technology, in which 'no acts of God can be permitted' (Alfvén, 1972), we cannot call on experience, since the magnitude of the problems confronting us are unique in human history.

International experience at many stages of the nuclear fuel cycle supports the thesis that describing nuclear safety problems as 'amenable to engineering solution' confuses the way things are with the way one would like them to be. This experience shows that people have impressive talents in overcoming foolproof systems, and suggests that catastrophes have so far been averted more by luck than design (Lovins, 1973, p. 00).

P. Chapman, 'The ins and outs of nuclear power', *New Scientist*, vol. 64, no. 928, 19 December 1974, pp. 866–9.
A. V. Kneese, 'What will nuclear power really cost?' *Not Man Apart 3*, 5, 16 (May 1973), Friends of the Earth; based on testimony of the USAEC hearings on the nuclear fuel cycle, November 1972.
H. Alfvén, *Bulletin of Atom Science*, 28, 5, 5, May 1972.
A. B. Lovins, *World Energy Strategies*, Earth Resources Research for Friends of the Earth, 1973.

Nuclear mining. The use of controlled underground nuclear explosions to fracture rock too deep to mine, too inaccessible, too low grade or all three is one potential use of nuclear explosions. The ore would be leached by SOLUTION MINING techniques.

Underground gasification of poor quality and/or inaccessible coal seams is suggested. In this case two shafts would be drilled down to the coal seam, a controlled nuclear explosion used to shatter the seam, ignition of the coal would be initiated at the foot of one shaft and by controlling air flows down this shaft a stream of combustible gas would be obtained up the other shaft.

These proposals rest on the assumption that the explosion can be controlled and that radiation release to the environment is zero or minimal. Currently, there appears to be little interest in the commercial development of these techniques.

Nuclear power, Some ethical considerations. Unlike any other human product, nuclear wastes will survive for periods that are more appropriate to geology than to history. The HALF-LIFE of plutonium–239 is 24 400 years. For every kilogramme of plutonium created today, there will still be 500 g left in 24 400 years, 250 g in 48 800 years, 125 g in 97 600 years, and so on. In view of the extreme toxicity of such a material, we cannot ignore serious doubts about our right to saddle our descendants with it – particularly as we have, as yet, devised no practical means of disposing of it. We have entered into a Faustian bargain whereby we are given an unlimited energy source in return for a pledge of eternal vigilance (Edsall, 1974).

We have no means of knowing whether the safety problems we are setting ourselves are capable of solution. The nuclear industry rejects but cannot refute this thesis. It points with justification to the very strict controls and high levels of safety within the industry. But such statements avoid rather than answer the central question of whether the problems of plant safety and of containment and disposal that we are setting ourselves are inherently insoluble. Unfortunately this dispute is not essentially a technical one and is not therefore resolvable on its technical merits.

People have to operate nuclear power-plants, no matter how much automation we introduce. People are forgetful; often they are irresponsible; and quite a few of them suffer from deep-seated irrational tendencies to hostility and violence. . . . I believe that the confident advocates of the safety of nuclear power-plants base their confidence too narrowly on the safety that is possible to achieve under the most favourable circumstances, over a limited period of time, with a corps of highly trained and dedicated personnel. If we take a larger view of human nature and history, I believe that we can never expect such conditions to persist over centuries, much less over millennia (Edsall, 1974).

We can only hope that 'the safety of the public . . . will never be made dependent upon almost superhuman engineering and operational qualities' (Cottrell, 1974).

A. Cottrell, Letter to the *Financial Times*, 7 January 1974.
J. T. Edsall, *Environmental Conservation*, vol. 1, no. 1, 1974, p. 32.

Nuclear reactor designs

Burner reactors

This is essentially a pressure vessel contained in a biological shield to contain radiation release. It has the means for inserting the fuel and removing the waste products; moderators for slowing down the neutrons from their fast velocities to thermal velocities so that fission can take place; control rods for controlling the overall speed of the nuclear reaction; coolant(s) for removing the energy liberated in the reaction; canning, i.e. containers for the fuel; and lastly the main constructional materials. The drawback with burner reactors (and they are virtually the only ones available commercially) is that the fuel is consumed and as uranium is a rare element and ^{235}U is less than 1 per cent of natural uranium, the burner reactors are considered a stop-gap measure until breeder reactors are available. Reactor designs vary in the fuels, moderators, coolants, canning, etc., and only the main designs are discussed.

1. Fuel can be natural uranium oxide or ^{235}U enriched uranium oxide. The enriched fuel has a much higher energy release per unit mass.

2. Moderators can be light (ordinary) or heavy water or graphite. (Heavy water is water containing a substantial portion of deuterium oxide, D_2O.)

3. Coolants come from light or heavy water, carbon dioxide, helium and sodium. (Lithium has been proposed for a possible fusion reactor.) Carbon dioxide is readily manufactured. Helium is obtained from natural gas but could eventually be in short supply. Sodium is used as a coolant in fast breeder reactors and is obtained by electrolysis of molten salt.

4. Cans – fuel containers in metallic and non-metallic varieties: Magnox (magnesium oxide alloy), stainless steel, zirconium, silicon carbide/graphite.

5. The materials of construction are steel, cement, copper, etc., and are the normal materials of conventional structures.

A summary of the common fission reactor types is given in Figures 40–6.

G. R. Bainbridge and A. A. Farmer, 'Nuclear reactors for the future', *Symposium: Energy Resources or Misuse?*, University of Newcastle upon Tyne, Institute of Chemical Engineering, September 1974.

Fast breeder reactors

The fast breeder reactor uses liquid sodium as a coolant, but because

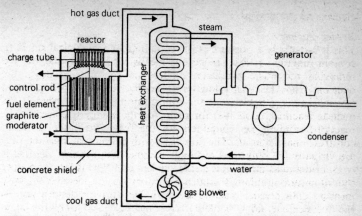

Figure 40. Magnox (magnesium oxide) reactor. The first British commercial reactor using metallic natural uranium metal fuel clad in magnesium alloy, Magnox, with graphite moderator and carbon dioxide coolant.

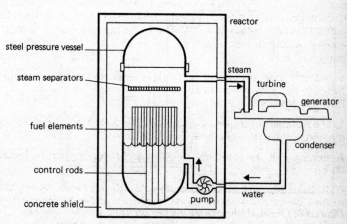

Figure 41. B W R. Boiling water reactor using enriched uranium oxide fuel pellets clad in Zircaloy, with light water moderator and coolant.

179

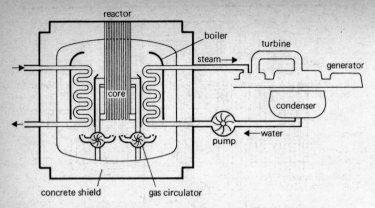

Figure 42. A G R. Advanced gas-cooled reactor using enriched uranium oxide fuel pellets clad in stainless steel, with graphite moderator and carbon dioxide coolant.

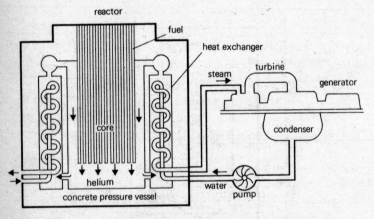

Figure 43. H T R. High temperature reactor using enriched uranium oxide or carbide fuel granules clad in carbon–silicon carbide, with graphite moderator and helium coolant.

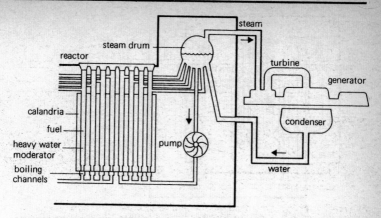

Figure 44. SGHWR. Steam generating heavy water reactor using enriched uranium oxide fuel pellets clad in Zircaloy, with heavy water moderator and light water coolant.

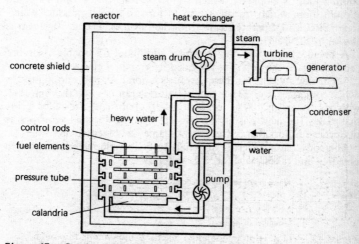

Figure 45. Candu reactor. Canadian reactor using natural uranium oxide fuel pellets clad in Zircaloy, with heavy water moderator and coolant.

181

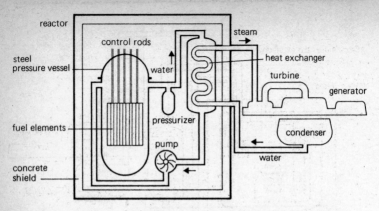

Figure 46. P W R. Pressurized water reactor using enriched uranium oxide fuel pellets clad in Zircaloy, with light water moderator and coolant.

of the inherent risks (however small) of a violent sodium–water reaction, special design techniques are employed to contain this reaction, if any, *outside* the reactor. Thus a primary liquid sodium cooling loop is employed which transfers heat to a secondary liquid sodium loop, which transfers heat in turn to a high pressure water-steam circuit to separate steam from the turbines. The arrangement is shown in Figure 47.

A breeder reactor is one in which additional fissionable fuels are produced from the neutrons provided by an initial charge of uranium–235. The raw material for these new fuels is non-fissionable uranium–238 which comprises over 99 per cent of natural uranium, and thorium–232, which is essentially 100 per cent natural thorium. They are called fertile materials. The neutrons are not slowed down to thermal velocities as in burner reactors, hence the generic name *fast* breeder reactors.

The conversion of fertile materials to fissionable materials is called breeding and takes place for ^{238}U as follows:

$$\overset{\text{(number of protons and neutrons)}}{\underset{\text{(number of protons)}}{^{238}_{92}\text{U} + \text{n} \longrightarrow {}^{239}_{92}\text{U} \longrightarrow {}^{239}_{93}\text{Np} \longrightarrow {}^{239}_{94}\text{Pu}}}$$

i.e. uranium–238 absorbs a neutron and becomes uranium–239. This changes spontaneously through two short-lived radioactive transformations to neptunium–239 and thence to plutonium–239.

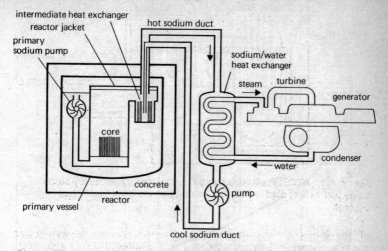

Figure 47. Fast breeder reactor with primary and secondary sodium circuits.

Similarly for thorium–232, which is transformed to thorium–233 which changes to protoactinium–233 and thence to uranium–233:

$$^{232}_{90}\text{Th} + n \longrightarrow \, ^{233}_{90}\text{Th} \longrightarrow \, ^{233}_{91}\text{Pa} \longrightarrow \, ^{233}_{92}\text{U}$$

Both plutonium–239 and uranium–233 are fissionable in an identical fashion to uranium–235. The thermal energy released per gramme of both materials is virtually the same as that of uranium–235. In a breeder reactor more fissionable material is produced than is consumed and in theory it is possible to convert all available fertile material, given that there is enough uranium–235 to start off with.

The breeding gain per charge currently achieved is about 1·2, and design improvements can get this up to 1·4. The doubling time is very important and currently about 20 years are required to double the initial amount of fuel. This could be shortened, but safety criteria may dictate otherwise. Prototype fast breeder reactors are now operating in France, the USSR and the UK and should lead to commercial designs within a decade. The need for this is evident in Figure 48, which shows the estimated 'world' uranium ore resources and demand using thermal reactors and the effect on availability of mining ore policies.

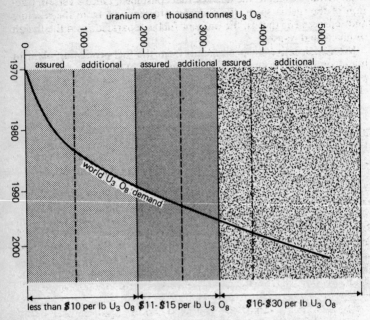

uranium ore   thousand tonnes $U_3 O_8$

Figure 48. Estimated world uranium ore resources and demand using thermal reactors. (From G. R. Bainbridge, A. A. Farmer and G. E. Shallcross, 'Materials resources for nuclear power', *Symposium on Energy Resources or Misuse?*, Institute of Chemical Engineering, University of Newcastle upon Tyne, September 1974.)

The picture is not comforting and should explain the crash programmes now under way in the UK and other countries for developing breeder reactors. It must be emphasized that burner reactors are very much a temporary measure to be phased out before the present supplies of uranium–235 are exhausted. The main constraints on development are fears as to safety and the availability of funds for further prototypes.

Nuclear reactor wastes. Radioactive wastes are produced by the controlled release of energy from fissile fuel and to a much lesser extent

184

by neutron bombardment of otherwise neutral materials in the reactor, including reactor metals, coolant fluids, air, carbon dioxide or other gases. The mass of radioactive fission products produced in a reactor is very nearly equal to the mass of fuel consumed. Once a certain fraction of the fuel is spent, the elements are removed from the reactors and reprocessed; that is, the unspent fuel is separated from the fission products and re-used.

Most radioactive waste emanates from the fuel-processing plant in liquid or slurry form and is stored in steel or concrete tanks for a preliminary cooling period prior to ultimate disposal. There are three classifications of radioactive waste slurries or liquids:

1. High level – radioactivity greater than 1 curie per gallon.

2. Intermediate level – radioactivity in the range 1×10^{-6} curie to 1 curie per gallon.

3. Low level – radioactivity less than 1×10^{-6} curie per gallon.

The problems associated with nuclear wastes are highlighted in the extreme case of PLUTONIUM.

Up to the mid-1970s in the UK about 30 tonnes of plutonium had been produced by the nuclear power programme. The planned expansion of fast breeder reactors would mean that this amount would increase to several hundred tonnes by the end of the century. Clearly these quantities represent a very substantial hazard and must be effectively isolated from the biological environment for many thousands of years.

Conventionally, radioactive waste is stored until natural decay processes reduce the activity to acceptable levels. For plutonium, the storage periods are so long that artificial storage is only a means of buying time until more satisfactory solutions, e.g. vitrification, are found. Vitrification is the convertion of wastes to a glassy end-product which is then sealed virtually for all time. Although the idea of vitrification of nuclear wastes has been around for ten years or more, little progress has been made beyond feasibility studies. Other proposals for high-level waste disposal include injecting them down deep boreholes into supposedly impermeable strata, but such ideas do not bear very close examination.

The guiding principles for any long-term programme of nuclear waste disposal were laid down by the US National Academy of Sciences in 1967 as:

1. All radioactive wastes should be isolated from the biological environment during their periods of harmfulness.

2. No waste-disposal practice, even if regarded as safe at an initially low level of waste production, should be initiated unless it would still be safe when the rate of waste production becomes orders of magnitude larger.

3. No compromise of safety in the interests of economy of waste disposal should be tolerated.

Currently high-level waste disposal is handled strictly in accordance with these three principles. However, both the first and the second principles are infringed by many present (UK and USA) practices involving the disposal of low-level liquid wastes into the sea and the release of gaseous wastes, after removal of most longer lived isotopes, through tall chimneys. While no one is being exposed to levels of radiation that approach anywhere near the maximum permissible dose, we may still wonder if these practices are sound should the nuclear power industry expand by a factor of 100 or more, as has been variously predicted for the twenty-first century. (◊IONIZING RADIATION, EFFECTS; IONIZING RADIATION, MAXIMUM PERMISSIBLE DOSE.)

Nuclear reactors, Classes. Plant in which a controlled chain reaction takes place. Currently all reactors are of the fission type and fall into two main classes – burners and breeders. (◊NUCLEAR REACTOR DESIGNS.)

Nucleus. The central core of an ATOM which is made up of two sets of particles, protons and neutrons.

Nutrients. The raw materials necessary for life which are consumed during the metabolic process of nutrition. Their type and consumption vary according to the particular plant or animal species. The main categories are proteins, carbohydrates, fats, inorganic salts (e.g. nitrates, phosphates), minerals (e.g. calcium, iron), and water.

Nutritional requirements, Human. Nutrients necessary for life and health in humans fall into five general categories: carbohydrates, fats, PROTEINS, vitamins and minerals. No one food contains all these nutrients, each of which can be found in a wide variety of foods. A completely satisfactory diet can be obtained by regularly eating foods from each of the following four categories:

1. Milk and dairy products, soybeans or whole small fish (for proteins, vitamins, calcium and other minerals).

2. Meat, fish, poultry and eggs (for protein, fats and vitamins).

3. Grains and starchy vegetables; e.g. wheat, rice, potatoes, corn (for carbohydrates, vitamins and some proteins).

4. Fruits and vegetables (for carbohydrates, vitamins, minerals and some protein).

Carbohydrates are required to provide the body with energy and may be converted into fat. Fats provide energy and may build up into reserves of body fat. Proteins provide material for growth and repair of body tissues. They also provide energy and can be converted into fat. Minerals are needed to provide the materials for growth and repair of the body and to regulate its processes. Vitamins are needed for the control of body processes.

The following table shows the energy value of certain foods expressed in terms of KILOCALORIES per ounce:

Butter	226	Beef	91	Apple	13
Cheese	120	Dates	70	Orange	10
Bacon	115	White bread	69	Turnip	5
Sugar	112	Potato	25	Lettuce	3
White flour	99	Milk	19		

O

Odour threshold. The concentration of an odour-bearing gas at which only half a panel of 'sniffers' can detect the smell. The odour is usually diluted in a dynamic system and presented to groups of volunteers at various dilutions. (◊ODOURS.)

Odours. The smell(s) produced by, usually, very small concentrations of organic vapours can produce violent aversion reactions in anyone exposed to them. The reactions range from nausea to insomnia. Many of these vapours come from operations such as animal by-products processes, farming, maggot breeding, brick-making and metallurgical processing. What is surprising is that the odour threshold in many cases occurs at very low concentrations of the substance in air (◊MERCAPTANS) and almost invariably these concentrations are well below the THRESHOLD LIMITING VALUE for the substance. For a 'normal' person, there is at least some comfort in knowing that mass poisoning is not taking place.

Nevertheless, odour suppressions at source can and should be practised on both public health and amenity grounds. The common methods are:

1. Absorption in a suitable liquid which may oxidize the offending vapour or neutralize it in the process (◊ABSORPTION).

2. Adsorption: ACTIVATED CARBON will adsorb organic molecules in preference to water vapour and can be tailored to provide optimum adsorption for the particular contaminant (◊ADSORPTION).

3. Incineration, i.e. oxidation at high temperature, will destroy all malodorous, gaseous and organic wastes but is usually very expensive.

4. After-burning of non-inflammable weak mixtures of organic vapours which heat the effluent air stream to temperatures greater than 750°C.

5. Ozonation: the use of OZONE's very powerful oxidizing abilities will deal effectively with some odours such as hydrogen sulphide.

In the UK, the Public Health Acts provide for enforcement of odour control. ($\Diamond$ODOUR THRESHOLD.)

Offshore oil extraction, Risks of. The extraction of offshore oil has five danger spots for large or catastrophic oil pollution:

1. Spill from the seabed or wellhead.

2. Pipeline fracture from the seabed to the oil rig.

3. Pipeline fracture or leak from the rig to the shore or a mooring buoy.

4. Spillage at a mooring buoy when tankers couple up and uncouple.

5. Collision with wellheads or rigs.

The result can be large OIL SLICKS which have substantial effects on fisheries, sea birds and coastal amenities.

Oil. Our major source of fossil energy resources at the present moment. Roughly two-thirds of the world's known ultimately recoverable reserves are in the Persian Gulf area, and the total effective lifetime of world reserves will probably be 70 to 80 years (Hubbert, 1969). It would appear that there is no geological justification for assuming that there will be further discoveries that will significantly alter the world picture. In any case, a doubling of world reserves would only delay depletion by a decade. Exploration experts predict that total world oil production will peak in the 1980s and decline thereafter. Our own North Sea reserves, once proven, represent only 1·5 to 1·8 per cent of ultimately recoverable world reserves (see Figure 49).

Those countries which are fortunate enough to have offshore reserves of oil are all planning massive investment programmes, to be followed inevitably by the most rapid possible rate of extraction – in the Norwegian sector, production is planned to peak in 1977, decline by half within four years and decline to almost zero by 1990 (Lovins, *World Energy Strategies*, p. 14). The longer-term social, environmental and economic arguments for a policy of restrained extraction carry little weight with governments concerned almost exclusively with what they foresee as occurring within their term of office and reinforced by the reasoning of DISCOUNTED CASH FLOW.

The current oil crisis appears to have taken the entire business and governmental world by surprise, in spite of all the evidence that has been readily available for many years now. In the USA a Cabinet Task Force on oil import control reported in February 1970 that 'we do not predict a substantial price rise in world oil markets over the coming decade', and in Britain, the Central Electricity Generating Board was assuming in the spring of 1973 that fuel costs would rise only a few

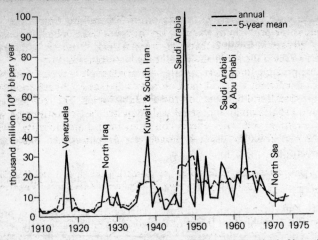

Figure 49. World oil discoveries (excluding Communist bloc). Note the importance of North Sea discoveries compared with Saudi Arabia. (From H. R. Warman, 'World energy prospects and North-Sea oil', *Coal and Energy Quarterly*, no. 6, Autumn 1975, p. 24.)

tens per cent to 1980 and remain constant thereafter (Lovins, *New Scientist*, 1973). And this despite closely argued cases, presented both by some persons within the oil industry and by OECD that oil prices would triple at least by 1980.

M. K. Hubbert, 'Energy resources', *Resources and Man*, W. H. Freeman, 1969.
A. B. Lovins, *World Energy Strategies*, Earth Resources Research for Friends of the Earth, 1973.
A. B. Lovins, *New Scientist*, vol. 58, 31 May 1973, p. 564.

Oil shales. Large oil-bearing shale deposits in the Green River formation of Colorado, Wyoming and Utah have been estimated to contain 1800 thousand million barrels of oil – more than four times the crude oil discovered to date in the USA. However, only 6 per cent of the deposits is accessible, yielding more than 30 gallons of oil per tonne of rock.

Pilot projects have demonstrated the feasibility of recovering shale oil (kerogen) and converting it to crude oil in liquid form. Commercial extraction by surface retorting is now possible (◊EXTRACTION OF OIL FROM SHALES).

Mining of the shale for commercial purposes must be carried out on a huge scale, however, in the region of 500 million tonnes per year. There is the problem of where to tip the spent shale, which is considerably greater in volume than the original rock. There is a danger of leaching from the spent shale polluting watercourses.

One possible limitation on oil-shale production is thought to be the lack of sufficient water in the Colorado watershed as there are already large agricultural demands on the available water. The recovery of oil from oil shales on an industrial scale requires vast quantities of water for cooling and the effects on the immediate locality, whose economic structure depends largely on irrigation, are likely to be considerable.

For more detail on the feasibility of extracting oil shales, see *US Energy Prospects – An Engineering Viewpoint*, US National Academy of Engineering, 1974.

Oil slick.   A floating layer of oil on the seas, rivers, canals or lakes, usually as a result of accidental spillage from tanks, pipelines, etc., or deliberate, as in the case of discharge of water ballast. Slicks can cause major local ecological disasters, especially to sea birds, and, if allowed to drift ashore, will foul beaches.

Common methods of eliminating slicks are:

1. To contain the oil slick by floating booms, and then transfer it into tanks by suction.

2. The oil can be absorbed on nylon 'fur' and then squeezed out of the fur for reuse or disposal.

3. Detergents can be used to break up the slick. However, this method has been attacked on the grounds that it does not remove the oil and detergents contain toxic ingredients harmful to birds and marine life, particularly shellfish.

Oligotrophic.   A lake that is oligotrophic is poor in producing organic matter – a characteristic of 'young' lakes and reservoirs. It is the opposite of EUTROPHIC.

Open-cast mining.   This the process of surface mining in which large quantities of mineral-bearing rock are scooped out to produce, in effect, a very large hole, with terraced sides. The largest, in Bingham Canyon, Utah, is $1\frac{1}{2}$ by $1\frac{2}{3}$ miles by $\frac{1}{2}$ mile deep.

The two basic types of open-cast mining are quarrying, in which the rock is cut into large blocks and usually transported away from the site in this form, and metal-mining, where the large quantities of rock are ground to a powder on site and the metal-bearing minerals are removed.

In this way large quantities of poor ores may be processed in order to obtain relatively small amounts of the valuable minerals. This method also leaves a very large amount of useless ground-up rock.

The scale of some of these operations is illustrated by the Anaconda Company's mine at Twin Butte near Tucson, Arizona, where 236 million tonnes of overburden and rock were removed to get to the low-grade copper ore 600–800 feet below ground. The ore itself has a copper content of 0·5 per cent, i.e. 200 tonnes ore are required to obtain 1 tonne copper.

Organochlorines. A major class of chemicals emanating from the organic chemicals industry. It includes CHLORINE, insecticides, AEROSOL PROPELLANTS, POLYCHLORINATED BIPHENYLS, PVC, DDT, DDVP, and ENDOSULFAN. Organochlorines are characterized by persistence, mobility and high biological activity. They have very long HALF-LIVES: that for DDT is ten years or more, for DDE many decades.

All organochlorines have a high capacity to injure living systems and allied with their other attributes may possibly constitute the greatest threat to our life-supporting ECOSYSTEMS and associated biological cycles.

In many insecticide applications the organochlorines are being replaced by the ORGANOPHOSPHATES and the new synthetic PYRE-THROIDS are likely to be produced at rates which might rise to 20000 tonnes annually world-wide. (◇ PESTICIDES.)

Organophosphates. A group of chemical pesticides which embraces such names as Azodrin, Malathion, Parathion, Diazinon, Trithion and Phosdrin. All are descendants of nerve-gas production in the Second World War and block the central nervous system by inactivating the enzyme responsible for breaking down a nerve 'transmitter' chemical called acetylcholine. The result is hyperactivity resulting in death.

Organophosphates are generally very much more toxic to insects than mammals and also have a very much shorter HALF-LIFE than the ORGANOCHLORINES. For these reasons they are labelled as 'safe' insecticides, although they may be CARCINOGENIC if precautions are not taken during use. However, the long-term ecological effects of these chemicals is not known. They have only been in use for one generation which is insignificant on the evolutionary time-scale.

In common with most similar products, these chemicals buy time which must be put to use for the stabilization of populations and resource consumption and planned ECOSYSTEM management. (◇ PES-TICIDES.)

Oxygen cycle.   Oxygen is a major component of all living matter and is vital in the free state for the higher animals which require it in their metabolism. Its presence on earth is almost certainly due to the process of PHOTOSYNTHESIS which is the assimilation of carbon dioxide and water for the production of carbohydrates and free oxygen. This free oxygen both supports and comes from life.

The emergence of free oxygen has its origin around 3000 million years ago when simple AUTOTROPHIC ORGANISMS evolved which were able to split water and release oxygen. (There is geological evidence of oxidized sediments around 1500 million years old in the form of ferric compounds, the oxidized form of ferrous rocks.)

With the emergence of free molecular oxygen (O_2), the sun's energy split the oxygen molecules into atomic oxygen (O), which is highly reactive and forms ozone (O_3) which has built up into the OZONE SHIELD. Thus the earth's atmosphere began to evolve and stabilize. However, oxygen produced by photosynthesis is used up by respiration either by the consumers in the FOOD CHAIN or the decomposers ($\lozenge$CARBON CYCLE). So, if the oxygen is recycled in this manner, how did atmospheric concentrations build up?

The answer lies in the carbonate and carbonaceous sediments; calcium carbonate (limestone) is an example of the former and coal an example of the latter. The sediments formed by the deposits of animal and plant bodies removed carbon from the carbon cycle and tied it up for geological time. For every atom of carbon laid down to sediment two atoms of oxygen are left free. Thus, the bank of free oxygen that we have in the atmosphere was made possible by the formation of carbonaceous sediments. As the sediments were being deposited, the atmosphere evolved, the ozone shield grew, and we now have a stable atmosphere. The biosphere and the atmosphere evolved simultaneously due to the carbon and oxygen cycles operating together.

If photosynthesis were to stop tomorrow the reservoir of oxygen would be sufficient to sustain higher life for millions of years without significant depletion.

The carbonaceous sediments numbers many thousand million (USA billion) tonnes, of which all the coal and oil likely to be used by man is much less than 1 per cent. Therefore, the combustion of coal and oil, while leading to an increase in carbon dioxide content, will not significantly affect the oxygen content, although it may, through the GREENHOUSE EFFECT, alter the earth's radiation balance.

The indivisibility of the biospheric processes is clearly illustrated in the oxygen cycle. We should take care not to abuse that which we do not understand and are unable to control.

Oxygen deficit. ▷DISSOLVED OXYGEN.

Oxygen demand. ▷DISSOLVED OXYGEN.

Oxygen sag curve. ▷DISSOLVED OXYGEN.

Ozone (O_3). Ozone contains three atoms of oxygen, whereas the atmospheric oxygen molecules contain two atoms (O_2). Ozone is used as an oxidizing agent, in water treatment for example (▷WATER SUPPLY). It is manufactured by electrical discharge in oxygen or air. It is also produced by ultra-violet radiation and forms the OZONE SHIELD for the earth.

Ozone is a dangerous irritant to eyes, throat and lungs. It can be formed in PHOTOCHEMICAL SMOG. Levels up to 0·9 part per million have been measured in Los Angeles. The first-alert level in Los Angeles is 0·5 part per million. The MAXIMUM ALLOWABLE CONCENTRATION for an eight-hour exposure is 0·05 part per million. It can also be a threat to aviation personnel as cabin air levels at supersonic flight, using ambient air under pressure, contain about 2·5 parts per million.

Ozone shield. A layer of OZONE surrounding the earth formed by ultra-violet radiation which splits molecular oxygen (O_2) to two atoms of oxygen which is highly reactive and forms ozone (O_3). The ozone shield acts as a barrier to the radiation and protects the BIOSPHERE. The maximum concentration is found between 15 and 30 kilometres from the earth's surface and it is at this height that supersonic aircraft are expected to fly. There is speculation that the shield may be substantially affected by this activity and by the effects of CHLORINE released by AEROSOL PROPELLANTS. If the shield is reduced, this may increase the incidence of radiation-induced skin cancer. Recent attention has also focused on nitric oxide (NO) which also attacks the layer. (▷OXYGEN CYCLE; NITROGEN OXIDES.)

P

Paper. A matrix of CELLULOSE fibres usually free of non-cellulosic materials. The main classes are:

Newsprint: made from mechanically ground wood PULP (which contains short cellulose fibres and some LIGNIN) and recycled newspapers. Up to 40 per cent of newsprint is recycled and this factor is expected to increase. (⟡DE-INKING; RECYCLING.)

Kraft: a strong brown paper with long cellulose fibres, usually made from a sulphate pulp. It is used for packaging and wrapping. It can be readily recycled.

Printing and writing paper: usually bleached, with a fine texture and containing special fillers for ink absorption. Recycling is usually easily accomplished.

Speciality: this embraces papers having a high strength when wet, papers treated with resins, paper towels, papers for photographic emulsions, etc. Recycling is often difficult as resin types vary; therefore separation at source is essential.

Parasites. Organisms that live attached to or in living organisms. They gain food and often shelter but the host gains nothing and usually suffers as a result.

Parathion. ⟡ORGANOPHOSPHATES.

Particulate pollutants, Control. ⟡ELECTROSTATIC PRECIPITATORS; CYCLONES; THRESHOLD LIMITING VALUE; THREE MINUTE MEAN CONCENTRATION.

Particle, Fundamental or elementary. This term, used in nuclear physics, refers to any particle of matter that is not composed of simpler units. The electron, proton and neutron, the main components of any atom, were the first to be discovered and researched. However, since then others have been identified, such as the positron, the neutrino, mesons and hyperons.

Electrons, protons, positrons and neutrinos are stable; the remainder, mesons, hyperons and neutrons, decay spontaneously into fundamental particles of a lower mass, accompanied by the liberation of energy. Neutrons will only decay spontaneously when isolated from the atomic nucleus. Mesons and hyperons, of which there are many different types with different properties, have mean half-lives of less than 0·000003 seconds.

Particulates. Fine solids or liquid droplets suspended in the air. The solids often provide extended surfaces due to their irregularities and therefore other pollutants can be carried along; for example, smoke particles and sulphur dioxide have greater effects on health when in combination than when emitted separately. It is postulated that the smoke particles are carried deeper into the respiratory tract with the sulphur dioxide 'attached' (or adsorbed) and thus the medical effects are compounded.

The term particulates as used in air pollution includes all the separate terms: grit, dust, fume, aerosol, smoke, etc.

PCBs (Polychlorinated biphenyls). Chlorinated hydrocarbons used as plasticizers and in transformer-cooling oils to enhance flame retardance and insulating properties. They are highly active biologically and their use is now (as a result of voluntary restrictions) mainly confined to closed-circuit applications where release to the environment is difficult.

They are present in seawater such as the Clyde estuary, where, in 1969, concentrations of less than 0·01 microgrammes per litre were found. Concentration in mussels was between 10 and 200 microgrammes per litre, which gives a concentration factor of between 1000 and 20000. They have been blamed, in conjunction with PESTICIDES, for the death of sea birds in times of stress. Seawater concentrations are now falling rapidly due to the voluntary limitations on their use by both manufacturers and users.

Pentosans ($C_5H_8O_4)_x$. Polysaccharides present with cellulose in plant tissue including straw and sawdust. They can be utilized by HYDRO-LYSIS which converts them to pentoses (general formula – $C_5H_{10}O_5$) for the production of FURFURAL.

Percolating filter. ⬦SEWAGE TREATMENT.

Peroxyacetylnitrate (PAN). One of a number of complex compounds present in PHOTOCHEMICAL SMOG. It causes irritation to eyes and is toxic to plants.

Pest control. The term for the control of pests (e.g. tsetse flies, mosquitoes, cotton bollweevils) which affect public health, or attack resources of use to man.

The techniques used other than PESTICIDES are listed below:

1. Sterilization: i.e. use of irradiated males which controls or stops reproduction of the species (⋄IRRADIATION).

2. Administration of juvenile hormones so that the species cannot metamorphose and therefore cannot reproduce.

3. Exposure to sex attractants: phemerones, specific chemical produced by the female attracts the male into an insect trap and hence the male can be destroyed.

4. Plant breeding to develop insect-resistant plants.

5. Modified planting practices:
   (a) a favourite food is planted nearby, therefore drawing the insect to it.
   (b) crop rotation.

6. Encouragement of predatory insects or animals: e.g. the use of small fish which prey on mosquito larvae.

7. Infection by viruses, bacteria or parasites: e.g. a strain of nematode (minute thread worm) which is a parasite of mosquitoes is being introduced for mosquito control.

8. Deprivation of breeding ground: e.g. drainage of swamps to deny mosquitoes their breeding grounds, controlled tipping of DOMESTIC REFUSE to deny breeding grounds to flies. (⋄INSECTS.)

Pesticides. A product or substance used in the control of pests such as vermin, mosquitoes, moulds, all of which may affect public health or attack resources of use to man. (The term subsumes insecticides with which it is often used interchangeably.)

There are three main classes:

1. Chlorinated hydrocarbons (e.g. DDT) which are long-lived and capable of being concentrated biologically.

2. ORGANOPHOSPHATES which are short-lived and degrade to 'harmless' end-products.

3. Artificial pyrethrums originally based on natural sources from pyrethrum flower heads but now being synthesized in very large amounts.

Widely used pesticides are listed below.

Chlorinated hydrocarbons (persistent in the environment)
DDT and its metabolites, e.g. DDE: used widely to control malarial mosquitoes and houseflies, but strains have evolved with resistance. Acute oral toxicity to mammals low, but it becomes concentrated in fatty tissues, especially around and in vital organs.
Aldrin and Dieldrin: formerly widely used as seed dressing, but use suspended because of death of birds and other wildlife. Their use is restricted in the UK because of effects on the FOOD CHAIN.
BHC and lindane (its gamma isomer): used for seed treatment. At one time widely used in gardening products.

Organophosphorus insecticides (short-lived in the environment)
Parathion and Malathion: extremely dangerous to use – related to nerve gases and lethal in very small doses.
Carbamate insecticides (also used in fungicides and herbicides): fairly short-lived; attack nerve function.

Pyrethrins (◇PYRETHROIDS)
Pyrethrin I, Allethrin, Bioallethrin: natural products or closely related to natural products. Often a mixture, making build-up of resistance in houseflies, for example, more difficult. Relatively safe.

The chlorinated hydrocarbon class concentrates in the fatty tissues and vital organs of birds. In time of stress, body fat is mobilized and death can result. A classic case was the Irish Sea bird wreck in 1969 when about 10000 sea birds died in a storm. Later analysis showed that body-fat mobilization carried PCBs to the vital organs and certainly contributed to the wreck.

It is postulated that pesticides have synergistic effects in combination with air pollution and that dietary deficiency can markedly increase any hazard, particularly with the ORGANOCHLORINE and ORGANO-PHOSPHATE groups.

Another class, not now often used, are the inorganics, which are preparations of zinc, copper, arsenic or mercury. All are extremely toxic to human and animal life, but *may* be of use should resistance to organics develop.

Pesticides have undoubtedly contributed greatly to human health and increased food yields and will continue to do so – but unless caution is exercised, man will be in a race to develop new and/or more powerful insecticides as resistant strains of pests develop. For dangers of overzealous insecticide use, ◇ECOSYSTEM. (◇CARCINOGENS; TERATOGENS.)

pH. A measure of the alkalinity or acidity of a substance. The pH value of any solution in water is expressed on a logarithmic scale to the base 10 ($\diamond$LOGARITHMS). It is defined and calculated as the reciprocal of the hydrogen-ion concentration of a solution and may be expressed in symbolic form as

$$pH = \log_{10}\left(\frac{1}{H^+}\right)$$

where H^+ is the concentration of hydrogen ions.

What this means in practice is that the pH scale ranges from 0 to 14 with the mid-point 7 indicating neutrality. If acid is added to water, the H^+ value increases and the pH decreases. Thus, a pH value less than 7 is acidic. If greater than 7, it is alkaline (the opposite of acidity). Each unit increase in pH value expresses a change of state of 10 times the preceding state (because of the logarithmic scale). Thus, pH 5 is 10 times more acidic than pH 6; and pH 9 is 10 times more alkaline than pH 8.

pH measurements can be made by observing colour changes in special indicator chemicals or with indicator impregnated paper (litmus paper), and also by pH electrodes. The pH value of effluents, toxic liquid wastes, etc., is a crucial parameter in effluent treatment. Acidic solutions, for example, would require neutralization with an alkali prior to disposal so that the pH is 7, and treatment can be undertaken without the added complications of acidity (or alkalinity).

The pH of soils is an important factor in soil management – acidic soils require neutralization with limestone (calcium carbonate) which replaces one hydrogen ion in the soil with one calcium ion. However, as the calcium is leached ($\diamond$LEACHING), it must be replaced or the acidity returns. Other techniques, such as the use of CHELATING AGENTS, may also be used in soil management to remove or sequester unwanted ions.

Phosdrin. $\diamond$ORGANOPHOSPHATES.

Phosphorus (P). An element that plays an essential role in the growth and development of both plants and animals. In plants it is the energy exchange between adenosine diphosphate and adenosine triphosphate that provides energy, derived as a by-product of photosynthesis, at sites removed from the green parts in which photosynthesis occurs. In animals and plants, phosphorus is an essential component of DNA. It is impossible to conceive therefore of any substitute for phosphorus as a plant nutrient, and a phosphorus deficiency will retard growth and seriously affect yields ($\diamond$PLANT NUTRITION).

In FERTILIZERS, phosphates may be associated with excessive quantities of fluorine, a cumulative plant poison.

Phosphorus cycle. The phosphorus cycle is sedimentary as are those of the elements calcium, iron, potassium, manganese, sodium and sulphur. In essence, the cycle operates as follows: compounds such as calcium phosphate, sodium phosphate, etc., are leached from the rocks into the soil and water, where they are taken up by plant roots. The plants are subsequently consumed by herbivores, which in turn are eaten by carnivores. On the death of either herbivores or carnivores, decomposition takes place and the compounds are returned to the soil through the action of water – hence sedimentary cycle.

In the sea there is a similar cycle in which phosphate compounds from sediments pass through the aquatic food chain, some fish being eaten by sea birds, whose droppings (guano) are rich in these compounds, and recycling eventually takes place.

Photochemical smog. Photochemical smog appears to be initiated by nitrogen dioxide. Absorbing the energy of sunlight, it forms nitric oxide to free atoms of oxygen (O), which then combine with molecular oxygen (O_2) to form ozone (O_3). In the presence of hydrocarbons (other than methane) and certain other organic compounds, a variety of chemical reactions takes place. Some 80 separate reactions have been identified or postulated. Many different substances are formed in sequence including formaldehyde, acrolein, PAN, etc. The low-volatility organic compounds formed condense and form a characteristic haze of minute droplets which is called photochemical smog. The organics irritate the eye, and also, together with ozone, can cause severe damage to leafy plants such as tobacco and endive.

It frequently occurs in the Los Angeles area and its persistence has resulted in the passing of legislation to drastically curb automobile emissions. It is not unknown in London and the industrial regions of Holland in the summertime when TEMPERATURE INVERSIONS are prevalent. (⟡AUTOMOBILE EMISSIONS; SMOG; NITROGEN OXIDES.)

Photosynthesis. The means by which CHLOROPHYLL enables radiant energy to be used to accomplish the chemical conversion of elements in the atmosphere into organic matter. Chlorophyll is contained in organisms such as green and purple bacteria, blue-green algae (in fresh water), phytoplankton (at sea) and green plants (on land). The organisms live in those areas that receive sunlight such as the top few centimetres of soil or rivers and lakes. In the seas sunlight can penetrate over 100 metres and this is the province of phytoplankton. On land the green plants account for most photosynthesis.

Photosynthesis can be summarized as:

$$nCO_2 + 2nH_2A + energy \longrightarrow (CH_2O)_n + nA_2 + nH_2O$$

or carbon dioxide plus hydrogen donor plus energy gives organic compounds (carbohydrates) plus a free, i.e. gaseous, compound plus water. Photosynthesis, as carried out by green plants and phytoplanktons, uses carbon dioxide plus water for the hydrogen donor and the equation for such a reaction is:

$$nCO_2 + 2nH_2O + energy \underset{\text{green plant}}{\longrightarrow} (CH_2O)_n + nO_2 + H_2O.$$

In this way the oxygen content of the atmosphere is maintained (⟡OXYGEN CYCLE), carbon dioxide is fixed (⟡CARBON CYCLE), and a carbohydrate source, $(CH_2O)_n$, is available for incorporation in cell structure or as a source of energy directly or indirectly for all plants and animals.

The 'purple' and 'green' bacteria can use hydrogen sulphide (H_2S) as the hydrogen donor with sulphur as a by-product.

Photosynthetic efficiency. The percentage of total energy falling on the earth that is fixed by plants. It is approximately 1 per cent.

Pica. The abnormal craving for unusual foods. A common example is found in children – usually in deprived urban areas where lead paint on railways, walls, etc., may be found – who eat the paint flakes because of the sweet taste. Lead poisoning then results from the ingestion of lead paint fragments.

Plant nutrition. The essential minerals for plant growth include phosphorus, potassium and calcium. In most tropical areas the soil cannot maintain adequate reserves of these minerals because of the heavy rainfall. Such soils also contain high levels of iron and aluminium oxides. In tropical jungles most of the nutrients are concentrated in the vegetation rather than the soil, and since such vegetation is evergreen, there is little chance for the nutrients to build up in the soil as they do in temperate deciduous forests.

When tropical forests are cleared, therefore, the thin soil is soon washed away by the heavy rainfall, leaving the iron and aluminium oxides, which eventually weather into a rock-like substance called laterite. Because of these effects, it is doubtful whether significant quantities of food will ever be obtained by clearing the vast areas of tropical forest that exist on earth.

Plastics. Any substance that is capable of plastic flow or deformation under certain conditions or at some stage of its manufacture and thus can be moulded into shapes by heat and/or pressure. The common definition of plastics relates to those products of the chemical industry called 'polymers' which fall into two groups, thermoplastics and thermosetting.

Thermoplastics retain their potential plasticity after manufacture and can be re-formed by heating. The main types in this group are polyethylene, polypropylene, polystyrene and polyvinyl chloride (PVC) which embrace the whole range of domestic use.

The thermosetting group includes mainly resins such as the Bakelite or epoxy varieties, which are made into light switches, etc., and cannot be reused as there is permanent and irreversible change in the chemistry on setting. (◊ PLASTICS, RECYCLING.)

Plastics, Degradation. The very durability of plastics which is an excellent property for many purposes makes them virtually indestructible when discarded, and as RECYCLING is difficult on both cost and quality grounds, means have been sought to make them degrade when their useful life is over. There are two postulated routes, photodegradation using ultra-violet radiation (solar radiation) and biodegradation.

Photodegradation makes use of the fact that window glass *removes* ultra-violet radiation. Therefore plastic goods kept indoors are not exposed to ultra-violet rays, but when discarded on tips, etc., they would be exposed. Thus, the incorporation into the polymer of ultra-violet-sensitive groups would cause degradation on rejection.

Biodegradation is virtually impossible as no commercial plastics are biodegradable. The only real solution is to use less plastics for non-essential purposes.

Plastics, Recycling. In 1971 the UK plastics production was 1500 thousand tonnes, of which 25 per cent was for packaging, 8 per cent for housewares and toys, 26 per cent for building materials, 20 per cent consumer products and furniture, and 7 per cent for transport uses. Thus, only 33 per cent of production has a very short life – the rest are virtually in captive use. This 33 per cent are all thermoplastics and can in theory be recycled (◊ RECYCLING). However, direct recycling to obtain virgin plastics is virtually impossible due to contamination, mixtures of grades, and the type of original product.

An indirect recycling route is to take virtually as received plastics waste and to grind it, mix in fillers and put it through a high energy extruder and use the resultant material for 'low quality' purposes such as fence posts, pallets and roof tiles, where finish is relatively unimportant but the plastics property of durability and resistance to decay is. This method has promise if plastics waste can be collected in sufficient quantity. However, unless waste plastics can be collected at source, seperation from DOMESTIC REFUSE is not economic as it comprises less than 2 per cent by weight. Thus the normal domestic refuse disposal processes will prevail, but with the proviso that PYROLYSIS would be the most suitable energy recovery process for plastics because of their hydrocarbon content and high volatility.

F. Flintoff, *Plastics in the Environment*, British Plastics Federation, 1973.

Plume rise.   The rise of plumes from chimneys is a function of atmospheric conditions and the plume discharge (efflux) velocity, temperature and density. In still atmospheric conditions the plume will rise vertically; in strong WIND conditions it can be carried away horizontally or drawn downwards in the low pressure area behind the chimney. Further complications arise if the chimney is in a built-up area or in countryside of varying contours as the wind patterns can then cause downdraughts much more frequently.

To avoid downdraught, the exit velocity of the plume should be at least one and a half times the wind velocity. Thus, where noxious fumes are concerned, a meteorological survey is required to determine the annual wind velocity profile so that the plume will not be drawn down for, say, at least 95 per cent of the year. The stack gas temperature is also very important as this determines the gas buoyancy. The higher the temperature, the greater the buoyancy. The gas eventually cools to ambient and then the plume disperses downwind in an ever-widening cone.

Plutonium (Pu).   Plutonium–239 is an artificial radio isotope with a half-life of 24 400 years. It is made by bombarding uranium–238 with neutrons. It is reactive and emits ALPHA RADIATION of high penetrating power. It is a bone-seeking poison similar to radium but several times as toxic, and is one of the most hazardous substances known. Its most hazardous form may be the respirable plutonium dioxide particles produced by combustion – of the order of 10 thousand million particles per gramme of metal – with each particle suspected of carrying a substantial risk (perhaps 1 per cent or more) of lung cancer. The plutonium in breeder reactor fuel is already oxidized into a refractory ceramic which, it is claimed, cannot produce respirable particles. However, the sodium coolant in a breeder may be reactive enough in an accident to reduce the plutonium dioxide fuel back to plutonium metal. (◇NUCLEAR POWER; ◈PLUTONIUM, THE HOT-SPOT CONTROVERSY.)

A. B. Lovins, *World Energy Strategies*, Earth Resources Research for Friends of the Earth, 1973.

Plutonium, The hot-spot controversy. The magnitude of a dose of ionizing radiation is defined as the energy absorbed from the radiation per unit mass of tissue. If a specified amount of a radioactive material undergoes decay in an organ, the energy released can be calculated from a knowledge of the physics of the isotope in question. The fraction of this energy absorbed within the organ can be estimated from a knowledge of the penetration capacity of the relevant radiation: for example, the energy of α-rays is entirely absorbed within less than a millimetre from the source, whereas x-rays are much more penetrating and only a small fraction of their energy will be absorbed within a metre or so of their source. Such considerations provide an estimate of the total energy absorbed within the organ. The dose to the organ is then estimated by dividing that amount of energy by the mass of the organ. In general, the risk of cancer is considered to increase in proportion to the dose of radiation received by any tissue.

The hot-spot controversy is about the potential for lung-cancer induction by very small particles of plutonium that may be inhaled and lodge in the lung. Plutonium emits α-rays and it is contended by critics of present recommendations as to 'acceptable' doses that the conventional form of calculation may lead to gross underestimates of the dangers of such particles. They point out that all the energy of the α-rays emitted by the particle is absorbed in a small sphere of tissue surrounding the 'hot spot'. This sphere is no more than a millimetre or so in diameter and the dose within is therefore thousands of times higher than is calculated by averaging the energy over the entire mass of the lung. Cells immediately adjacent to the 'hot particle' will simply be killed, which is of little importance, but doses of *all* lower levels down to zero will be delivered at increasing distances from the particle. Thus, any particular dose critical in causing cancer must occur somewhere in the vicinity of the hot spot.

This argument is not accepted by agencies such as the International Commission for Radiological Protection. No detailed refutation has been offered, nor any fallacy or mistake pointed out. The controversy is likely to continue.

PNdB (perceived noise decibels). A frequency-weighted noise unit used for AIRCRAFT NOISE measurement. (◊ NOISE; DECIBELS.)

Pollutant. A substance or effect which adversely alters the environment by changing the growth rate of species, interferes with the food chain, is toxic, or interferes with health, comfort, amenities, or property values of people. Generally pollutants are introduced into the environment in significant amounts in the form of sewage, waste, accidental discharge, or as a by-product of a manufacturing process or other human activity. A polluting substance can be a solid, semi-solid, liquid, gas or sub-molecular particle. A polluting effect is normally some kind of waste energy such as heat, noise or vibration.

Pollutants may be classified by various criteria:

1. Natural or synthetic: sulphur dioxide is an example of a natural pollutant; the class of CHLORINATED HYDROCARBONS is an example of a synthetic pollutant. Natural substances can be assimilated into biological cycles; they may often undergo BIODEGRADATION. Synthetic substances, such as DDT, are not biodegradable; they are often toxic and accumulate in biological systems.

2. Effect: a pollutant may affect man, a complete ecosystem, a single individual member or component of an ecosystem, an organ within the individual member, a biochemical or cellular subsystem (e.g. crop growth rates).

3. Properties, e.g. toxicity, persistence, mobility, biological properties.

4. Controllability – the ease with which a pollutant can be removed from air or water is a very important factor. For example, most grit can be readily removed from flue gases, whereas sulphur dioxide cannot without a great deal of expense.

In addition, the environmental attributes of the system into which a pollutant is to be discharged must be taken into account. If a water-course is to be used for sewage discharge, the BIOCHEMICAL OXYGEN DEMAND imposed by the sewage must be such that it does not swamp the DISSOLVED OXYGEN of the stream.

The maintenance of biological processes is fundamental to our continued existence and health, and must not be grossly overloaded or resources can be irretrievably lost. The gross pollution of the Great Lakes is one example of how man has severely diminished a major natural resource.

For a discussion of the effects of pollutants, see M. W. Holdgate, 'The biological effects of chemical substances', *Chemistry and the Needs of Society*, Special Publication No. 26, The Chemical Society, 1974.

Polluter-must-pay principle. After the Second World War, when problems of pollution began to attract public attention and provoke protests, it was clear that the installation of plant and processes for pollution abatement would require financing from some source or other. Firms were evidently willing to install such plant if funds were provided from public sources, but such a policy did not have much appeal to taxpayers. Accordingly, much was made of the slogan 'the polluter must pay', interpreted as meaning that public money would not be used to subsidize pollution abatement for private industry.

However, in at least one case known to the authors this principle has been interpreted by the Department of the Environment in a peculiarly perverse manner. The Department was approached with a request to finance an investigation of the emissions from an industrial plant which were thought to be causing extensive damage to crops and property and to pose threats to health. The response was that, according to the principle that the polluter must pay, it would be improper for public money to be spent on such an investigation. (▷RIGHT TO DISCHARGE.)

Pollution.   An adverse alteration to the environment by a POLLUTANT.

F. Warner, 'Possibilities in pollution control', *Proceedings of the Institute of Mechanical Engineers*, vol. 187, 1973, pp. 115–27.

Pollution control.   The term for administrative mechanisms for control *and* the various processes and devices available for reducing emissions of waste streams.

The administrative control is effected by legislation (e.g. Alkali Act 1906, Clean Air Act 1968, Control of Pollution Act 1974, Health and Safety at Work Act 1974, and Radiological Protection Act 1970) and its enforcement and implementation through statutory bodies such as the Waste Disposal Authority for a region; the Water Resources Board; the Alkali Inspectorate for scheduled processes; and the Health and Safety Commission.

The actual control of pollution is through process selection and plant construction. For example, in stack gas emission control, is it best to filter out particulates and scrub and neutralize the gases; or specify chimney height and efflux velocity and rely on the atmosphere for dilution and dispersion? These are typical considerations for only one problem – dilute and disperse versus straight removal of the pollutant(s).

Pollution conversion.   In the elimination of one or more sources of pollution, it is important that new ones are not created. If solid waste is incinerated, air pollution may occur instead, which may be more serious than the original problem. Similarly, disposal of sewage SLUDGE too liberally on agricultural land could result in the build-up of toxic metals. Washing or scrubbing of exhaust gases can lead to a water-pollution problem.

Pollution indicators, Natural.   ▷BIOTIC INDEX; MOSSES; GLADIOLI; LICHENS.

Polymer. A chemical compound made by the repeated joining of MONOMER molecules. ($\diamond$ POLYMERIZATION.)

Polymerization. The joining together of MONOMER molecules by 'addition polymerization' in which case the POLYMER is a simple multiple of the monomer molecule, or by 'condensation polymerization', where the resulting polymer does not have the same empirical formula as the basic monomer constituent. The term is also used to cover the process of copolymerization, in which the polymer is built up from two or more different kinds of monomer molecules. Many plastics and textile fibres are made from natural or synthetic polymeric substances.

Population. Number of individuals of a certain species that live in a particular area at a particular time.

Population cycles. Regular patterns of change in a population over a period of time. They are most common in the simpler ecosystems of the world, e.g. tundra. The number of certain animals rises for several years and then drops sharply due to the delay in build-up of the controlling predator population, who then decimate the first population and consequently have no other food source to turn to. The predators in turn die out and leave the first population free to increase again. The cycle is affected by food supply, weather, disease, and competition.

Population growth. There are 3500 million people currently inhabiting the planet earth, and this number is increased by 70 million every year. It is the sheer size of the human race, together with its rate of growth, which is probably the most frightening and intractable factor of the 'environmental crisis'.

Behaviour which was acceptable when human numbers were much smaller is now potentially disastrous because, for the first time, the numbers of people on earth are sufficient, when combined with our technological skills, to inflict serious damage to our planet and its other inhabitants *on a global scale.* Furthermore, even if we could assume no further increase in the size of the human race (and of course we cannot), there is no conceivable hope of the majority of the people *at present* inhabiting the earth ever attaining anything approaching the standard of living of anyone capable of buying and reading this book – the earth's resources are already too limited to allow them to be equitably shared with our less fortunate fellows without 'breaking the global resource bank'. It is for this reason that Paul Ehrlich has called the underdeveloped countries 'never-to-be-developed countries'. Yet the populations of most of these countries are doubling every 20 to 30 years. The problem is exacerbated by the ubiquity of modern communi-

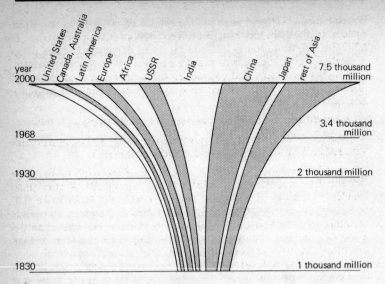

Figure 50. The world 'population tree'.

cations which make the people of these countries *aware* of the disparity between their own state and that of the people in the 'developed' (overdeveloped?) world. Not unnaturally they want a share. This combination of a rising population, with rising expectations and plummeting prospects is a potentially explosive situation.

Overpopulation in the developed countries takes a different form. While birth rates may, in some developed countries, become static or even drop slightly, expectations continue to rise. Yet even our present standard of living is based on a constant flow of protein *from* the underdeveloped countries. There can be no doubt that economic and industrial growth, like population growth, cannot be sustained in perpetuity.

That industrial expansion must either stop at a sustainable level or face the certainty of a precipitous drop is an exact parallel of the population explosion with its risks of famine and plague. In the industrial countries we replace labour with capital. We are likely to run out of food for our machines, or be overwhelmed by their ordure (Hussey, 1971).

(⟡EXPONENTIAL CURVE.)

M. J. L. Hussey, 'Has the twentieth century the technology it deserves?', *Institute of the Royal Society of Arts*, vol. 120, no. 5185, December 1971.

Precipitation. General term for the release of water from the atmosphere. It can be in the form of rain, snow, hail, dew or hoar frost. In all cases, condensation is initiated by nuclei of water droplets or ice crystals forming in clouds of moist air cooled below the DEW POINT. The nuclei then grow and coalesce in the clouds and then, provided the droplets are large enough to overcome any rising air currents, precipitation may take place.

Primary efficiency. ◊PHOTOSYNTHETIC EFFICIENCY.

Producer. Green plants that transform solar energy and carbon dioxide into sugars by PHOTOSYNTHESIS. (◊FOOD CHAINS.)

'Project Independence'. In 1973, as a result of the Arab oil-embargo against the United States, President Nixon announced 'Project Independence', a series of plans and goals to enable the USA to be self-sufficient in energy resources by 1980. This declaration was made without competent technical advice and is unrealizable. Subsequently, however, the US National Academy of Engineering set up a Task Force to consider the feasibility of such a programme by 1985.

The Task Force reported that the gap between energy production and energy requirements in the USA will continue to grow in the next ten years. At present the gap is filled by imports of oil and gas, and it is unlikely that demand can be reduced or supplies increased by a significant amount. Various steps could be taken to ameliorate the situation: economic incentives to industry to increase domestic supplies and implement new projects; government subsidies on relatively costly short-term measures such as coal liquefaction and oil-shale extraction (◊OIL SHALES); reduction in domestic demand by appeals to the 'conservation ethic', higher prices, smaller cars, better home insulation, etc.

Even if all these suggestsions were carried out, the USA would not be self-sufficient by 1985, nor might that be desirable in terms of world trade. The government must look to long-term solutions requiring massive support of research and development, both for its own economic good and for the sake of world-wide resources.

Despite the need for urgency stressed by the Task Force, neither the US Administration nor Congress have responded. At present the USA uses half of the total energy consumed in the OECD area, and it looks as though it is likely to continue as a profligate consumer.

US Energy Prospects – An Engineering Point of View, US National Academy of Engineering, 1974.

Proteins. Extremely complex organic compounds of large molecular weight which consist of hundreds of amino-acid molecules linked together. They are synthesized from AMINO ACIDS by and are present in all living things. They can also be manufactured from a variety of substances that include oil, cellulose, natural gas, sewage, agricultural wastes, through the action of yeast, algae, fungi or bacteria. The proteins are then separated from the substance, processed and used for human or animal feedstuffs.

Proteins are essential ingredients in the diet of all human and animal life. Individual protein requirements, like KILOCALORIE requirements, vary with body weight and activity. In addition to the calorific value, the quality of a diet can also be measured by the quality of the protein sources (◊NUTRITIONAL REQUIREMENTS, HUMAN). Where foods from animal sources are scarce, more protein is needed to compensate for the lower quality of the protein in vegetable sources. Protein deficiency is equally, if not more, serious as a deficiency in the total kilocalorie intake.

The daily per capita total protein supplies in 43 countries are shown in Figure 51.

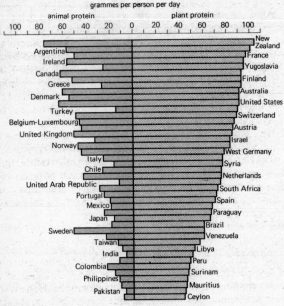

Figure 51. Daily per capita total protein supplies in forty-three countries. (After H. H. Cole, *Introduction to Livestock Production*, W. H. Freeman, 1966.)

Protein, World distribution of. It is a common belief that the industrialized nations share their plenty with less fortunate countries. It is true that some 2·5 million tonnes of plant-protein are annually supplied to the underdeveloped countries as cereals, but there is another side to the picture:

Outside the area of cereal grains most food and animal feed of the world market moves between the well fed and, still more surprisingly, from the hungry to the rich countries. This is particularly conspicuous in regard to protein, with the well-fed countries on balance making a net gain exceeding one million tons. Western Europe is buyer number one, followed by the United States; Japan dominates Asian trade in a similar way. . . .

Close to half of the protein taken from the Pacific (1966–68) emanates from . . . new fisheries in Peru and Chile – quite a remarkable achievement. Practically the entire catch is converted to animal feed in fish-meal factories. A major part (80 per cent) of the fish riches of the African Benguela Current is in a similar way channeled into reduction. . . .

In both cases almost the entire output of meal and oil goes to Western Europe (approximately two-thirds) and the United States (slightly less than one-third) with a minor part to Japan, where it constitutes a significant base for the budding animal production. The oil serves the European margarine industry. The impressive increase of animal production in post-war Europe is to a substantial degree due to this gigantic influx of first-rate animal protein, which enters into agricultural production almost through the back door . . . the two most protein-needy continents, Africa and South America, are the main suppliers of the largest quantities of animal protein feed moving in the world trade – and they provide those who already have plenty. . . . (Georg Borgstrom, *Too Many*, Macmillan Company, 1969.)

(◇ANCHOVY FISHERIES; FOOD FROM THE SEA.)

Protozoa. Small unicellular animals having a well-defined nucleus (in contrast to bacteria). They are found in a variety of habitats, eg., fresh and salt water and soil, and occupy an important position in these ECOSYSTEMS where they normally consume dead organic matter and wastes, although they can also be parasitic. (◇ACTIVATED SLUDGE.)

Provisional tolerable weekly intake. ◇MERCURY.

Public health. Public health embraces the health of both the individual and the community. In its widest sense it means the mental and physical health of the people, which in turn ensure the well-being of future generations – a resource to be husbanded just as much as the resources of land, air and water. All environmental factors are involved: food (type, quantity, wholesomeness), condition of work, home and play, pollutants, noise, etc.

Originally public health was primarily concerned with the prevention of diseases, e.g. CHOLERA or other sewage-borne diseases that used to abound in the UK and still do in many countries. We are still dependent on these early concepts and practices to prevent the return of such diseases, but also the wider spheres outlined above have great importance once basic health is established.

Public health embraces an awareness of new hazards, risks or pollutants introduced by technology in the name of progress. These hazards or risks must be controlled so that the benefit is maximized to all. We also learn how pollutants or substances (e.g. VINYL CHLORIDE) are more dangerous than first realized. Thus the field is dynamic and only through greater awareness of the risks of new substances or developments can progress be made. The aim of many of the entries in this book is to bring this awareness to the fore, so that we may use our new-found chemicals, process and practices wisely for the individual, the community and future generations.

The basic definition of public health has been given by C. E. A. Winslow, *The Cost of Sickness and the Price of Health*, World Health Organization, Monograph Series No. 7, 1951.

Pulmonary irritants. A group of air pollutants which affect the mucous lining of the respiratory tract which comprises the nasopharynx, the tracheobronchial area and the lung tissue or alveoli.

The nasopharynx defence mechanisms include the hairs at the nasal entrance, which filter out larger particles, and the mucous glands, which wash out many of those particles that escape the hairs. The bronchial tree contracts to prevent dust entry (e.g. coal dust) and hence reduce the amount of particulate matter. Coughing also ejects mucus in which the particulates are trapped. While bronchial spasms lessen the amount of dust, they also reduce the amount of air received. The alveoli consist of minute air sacs (around 300 million) filled with capillaries through which oxygen enters the blood and carbon dioxide is removed.

The main respiratory diseases are:

1. Emphysema, where the alveoli lose their oxygen/carbon dioxide exchange properties caused by smoking and chronic exposure to air pollutants.

2. Pulmonary fibrosis. Scarring of the lung tissue.

3. Pulmonary oedema. A drowning of the lung tissue in fluid caused by exposure to highly concentrated amounts of irritant or corrosive pollutants.

Severe pulmonary irritants include NITROGEN OXIDES, SULPHUR OXIDES and chlorine which can cause severe bronchial problems.

See G. L. Waldbott, *Health Effects of Environmental Pollutants*, C. V. Mosby, St Louis, 1973.

Pulp. The raw material for PAPER making. Pulp is usually obtained from trees, esparto grass and other long-fibred, CELLULOSE-containing materials. There are two main classes of pulp:

1. Chemical pulp, obtained from wood by chemical means, i.e. by sulphite, sulphate or soda processes which dissolve the LIGNIN and release the long cellulose fibres.

2. Groundwood or mechanical pulp, made by grinding the wood so that the fibres are separated. Groundwood is mainly used for newsprint because of its low quality, due to short fibre length.

The effluent from pulpmills has a high BIOCHEMICAL OXYGEN DEMAND and suspended solids and requires treatment before discharge. One recycling and pollution abatement method in certain pulpmill effluents is the manufacture of single cell protein by growing and harvesting yeasts on the dilute sugar solutions in the effluent.

PVC (Polyvinyl chloride). A plastic formed from the copolymerization of VINYL CHLORIDE ($CH_2 = CHCl$). PVC is widely used for covering electrical cables, rigid pipes, containers, etc. Its manufacture is based on the direct halogenation of ethylene at high temperatures to produce the monomer vinyl chloride. When disposed of by combustion (e.g. as litter in domestic refuse), it releases hydrochloric acid gas which can corrode incinerators rapidly.

The powder is a potential health hazard and is reported to be a cause of pneumoconiosis. It is also asserted to be as biologically active as ASBESTOS ($\diamond$HAEMOLYSIS TESTS). The degree of risk depends on particle size: the smaller the particle, the easier it is for it to penetrate the lungs. Particularly suspect are particles in the 1 to 5 micron range.

Pyrethroids. These are the active insecticidal constituents of pyrethrum flowers (*Chrysanthemum cinerariaefolium* and *Chrysanthemum coccineum*). The insecticidal properties are due to five differing compounds: pyrethrins I and II, cinerins I and II and jasmolin II. They are present in the achenes (single-seeded fruits) of the flowers in concentrations of 0·7 to 3 per cent, and may be extracted by organic solvents. As the cost of this extraction from the flowers is high, synthetic derivatives called allethrins are in common use.

Pyrethroids are highly unstable to the action of air, moisture and alkalis. The residues deteriorate rapidly after application. The characteristic insecticidal action is a very rapid knock-down of insect life followed by a substantial recovery due to detoxication enzymes present in the organisms.

Pyrethroids have low mammalian toxicity because of rapid detoxification by enzymes in the body.

R. White-Stevens (ed.), *Pesticides in the Environment*, Dekker, 1971, vol. 1, part 1.

Pyrolysis. The heating of organic matter or wastes containing organic matter such as DOMESTIC REFUSE in a closed retort in the absence of air. The subsequent volatilization produces combustible gases, a low-calorific value combustible char, a mixture of oils and liquid effluent (see Figure 52).

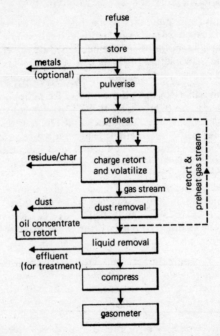

Figure 52. Flow diagram for pyrolysis of refuse for gas production.

The gas has a calorific value half that of natural gas and requires modified appliances for its combustion. The oil may not be present in sufficient quantities to justify a refinery stream in the UK and is therefore sent back for gasification in the pyrolysis reactor. The char can be upgraded to a fuel equivalent to low-grade COAL. The liquid effluent is treated to prevent water pollution. The process requires around 50 per cent of the fuel produced – the remainder is available for sale.

Pyrolysis is an embryonic refuse-disposal process which is receiving much attention in the USA as a means of conserving energy resources and recyling organic wastes.

R

Rad. ◊IONIZING RADIATION, DOSE MEASUREMENT.

Radiation. ◊IONIZING RADIATION.

Radioactive decay. ◊RADIONUCLIDE.

Radioactive half-life. ◊HALF-LIFE.

Radioactive isotope. The nucelus of an atom is made up of neutrons and protons. The number of protons in the nucelus is called the atomic number and characterizes the chemical element. The total number of neutrons (n) and protons (p) determines the mass of the nucleus. If two atoms have nuclei with the same number of protons but differing numbers of neutrons, they are said to be atoms of different isotopes of the same chemical element, and are shown with their respective mass numbers (n + p), e.g. ^{235}U and ^{238}U are isotopes of uranium. The isotopes of an element are identical in chemical properties and in all physical properties except those dependent on atomic mass.

Nuclear changes may result from neutron bombardment which transmutes an atom of one element to either an atom of another element or an atom of a different isotope of the same element.

For the use of the isotopes of uranium, ◊NUCLEAR ENERGY. (◊ISOTOPE; IONIZING RADIATION; HALF-LIFE.)

Radiological half-life. ◊HALF-LIFE.

Radionuclide. The nuclide of an atom that is radioactive, i.e. an atom that has an unstable nucleus which spontaneously disintegrates and emits ALPHA and BETA PARTICLES or gamma radiation, or both. Radium, uranium and strontium are examples of radionuclides, although radioactive isotopes of all common elements – carbon, potassium and hydrogen, for example – occur naturally. The process of spontaneous transmutation is called radioactive decay. (◊HALF-LIFE; IONIZING RADIATION; RADIOACTIVE ISOTOPE.)

Radon. A radioactive gas emitted naturally from rocks and minerals where radioactive elements are present. It is released in non-coal mines, e.g. tin, iron, fluorspar, uranium. Radon is an ALPHA PARTICLE emitter as are its daughter or decay products and has been indicted as a cause of excessive occurrences of lung cancer in uranium miners. New exposure standards have been adopted in the USA, Sweden and the UK based on a standardized dose measure called a 'working level month' or WLM. The new standards stipulate that a miner should not be exposed to more than 4 WLM per year.

Recessive genes. ◁IONIZING RADIATION, EFFECTS.

Recycling. Reusing a material not necessarily in its original form. The natural recycling of the substances required for life are the keystone or our existence on earth. In the CARBON CYCLE the reuse of the same material is obtained with the aid of solar energy.

We cannot by-pass the conservation laws of mass or energy and in a world of increasing material scarcity it is important to make the best use of all resources. The intelligent adoption of recycling techniques allied with good design practice which allows for materials to be reclaimed after their useful life is over can do much to conserve raw materials and energy, minimize pollution and save money. It would be too simplistic to expect people voluntarily to cut back their standard of living so that major energy and materials savings can be made, but recycling does offer the potential for considerable savings without major sacrifices on the part of the consumer.

Recycling falls into three classes.

1. *Reuse*: this is typified by the returnable bottle which makes several trips from bottler to consumer and back again where it is cleaned and refilled. Reuse may be allocated the highest availability in the recycling spectrum in that least energy and process complexity is normally expended in getting the material or article back into use.

2. *Direct recycling*: using the returnable bottle as our example, once it is unfit for reuse it may be cleaned and broken down to CULLETT at the glassworks and used to make more bottles. Direct recycling depends on the quality of the recycled material and its cost, which should not exceed that of the raw material. Currently most direct recycling occurs at the factory where the product is made, e.g. misshaped or broken bottles formed during glass manufacture are in fact fed back to the melting chamber. Industry calls this recycling and it is not to be confused with material reclaimed from waste or point of use. Thus paper with a 20 per cent recycled content may in fact be paper where surplus pulp fibres, mill offcuts and spoiled rolls have been internally re-routed back through the pulping process. Direct recycling has an intermediate

217

availability in that both energy expenditure and process complexity may be required in getting the material back into use.

3. *Indirect recycling*: this practice often makes no pretence at reclaiming the material for use as such but rather gets a second bite at the cherry. Continuing with our glass bottle, it is quite probable that it will eventually end up in domestic refuse where it can be extracted by screening and separation in conjunction with other bottles. These bottles will probably be of different colours and varying degrees of cleanliness and are unsuitable for cullett use unless costly optical sorting is used. (The reclamation of glass for remelting from refuse is more expensive in the UK because transport and collection costs are higher than the extraction of the basic raw materials. In the USA, however, optical grading of glass is under active consideration.) The bottles may, however, be ground up and used for a highly skid-resistant and durable road-surfacing material. Similarly waste plastic containers which *en masse* are unsuitable for direct conversion to new containers may be ground up and used for plastic fence-posts, pallets and chipboards, where appearance and structure are not primary considerations. Other forms of indirect recycling embrace the conversion of refuse to combustible gases, or the use of heat from the combustion of refuse for district heating by the means of incineration with heat recovery. Indirect recycling is the lowest form of recycling. Normally once processed in this phase, the material is no longer available for use except for landfill or incineration. The downgrading in use of several typical products is shown in Figure 53.

The recycling or recovery of useful material from domestic refuse is shown in the flowsheet of the Warren Spring Laboratory (UK) separation system in Figure 54.

Tipping is not a form of recycling – it is the sink for discarded materials, just as our surroundings form the last resting place for degraded energy. (◊LAWS OF THERMODYNAMICS; ENERGY.)

Recycling, Financial incentives for. A variety of taxes and levies could be imposed as incentives for recycling materials, and hence conserving resources, or to cover the costs of disposal of used materials.

Direct subsidies to reclaimers
Grants and price-support systems would ensure that recycled materials had the financial edge over original materials.

Disposal tax
A tax levies on a product according to the cost of disposal, e.g. a levy on a new car to allow for its eventual disposal. Suitable allowances

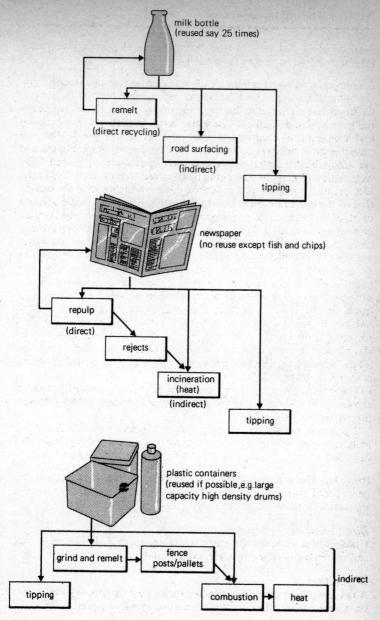

milk bottle
(reused say 25 times)

remelt
(direct recycling)

road surfacing
(indirect)

tipping

newspaper
(no reuse except fish and chips)

repulp
(direct)

rejects

incineration
(heat)
(indirect)

tipping

plastic containers
(reused if possible, e.g. large
capacity high density drums)

grind and remelt

fence
posts/pallets

tipping

combustion

heat

indirect

Figure 53. Recycling routes for various consumer products.

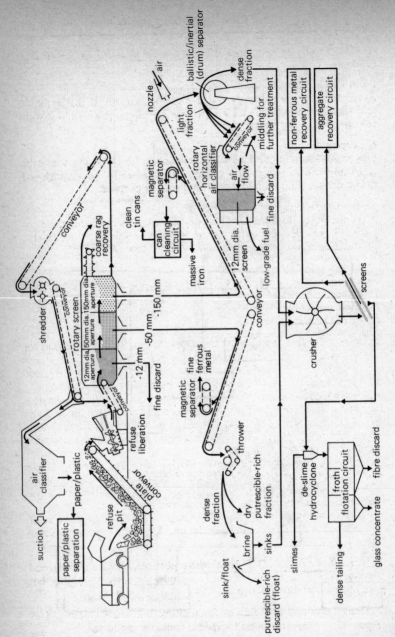

Figure 54. Warren Spring Laboratory — physical sorting of domestic refuse for the recovery of useful materials.

(i.e. deductions or avoidance of tax) can be included for recyclable products.

General pollution tax
A version of the POLLUTER-MUST-PAY PRINCIPLE. If one assumes that recycling processes give rise to less pollution, this would favour recycling.

Packaging tax
This to some extent overlaps with the disposal tax. Products would be taxed on the packaging value added or the amount of packaging used over a certain limit adjudged sufficient for the purpose.

Virgin materials tax
A tax applied to limit the use of particular materials by raising the price artificially. This would reflect their lifetime at current rates of use, e.g. for oil based on the RESERVES–PRODUCTION RATIO.

Virgin materials levy
Levies imposed on primary materials by producer countries to restrict consumption. Britain is likely to be on the receiving end of such a measure.
(◇COLLECTION TAX.)

Recycling, Product specification legislation for. This idea is based on the inherent wastefulness of the one-trip bottle and the throw-away-after-use syndrome. Legislation should be able to outlaw (in most cases) this wastefulness, and insist on a reusable or recyclable design. For example, the State of Minnesota has legislated for packaging controls which require all new packages and any changes in the packaging of existing products to be examined to see if they constitute a solid waste disposal problem or are inconsistent with the State's environmental policies.

Such legislation must be applied fairly; the glass, one-trip bottle should not be made the only banned form of container. Aluminium and plastics are much more energy intensive and thus they and other packaging materials must also be considered.

Product specification at the design stage can also achieve many of the desired ends of recycling legislation. The following criteria are suggested as being of use:

1. Increase the product's lifetime where possible, eliminating planned obsolescence. This should not only include increased durability but ease of repair and maintenance.

2. Design products for reuse or for multiple use where suitable. This

would mean in some cases the redesign of bottles to standard sizes and shapes with no non-standard moulding on the glass so that they can be used by any bottler.

3. Design for ease of reclamation and recycling. Where a product presents unnecessary problems in reclamation, it should be replaced by a more easily reclaimed product. This will depend on many factors, not the least the local conditions for reclamation. However, a number of products can be identified as difficult to reclaim and can be dealt with.

4. Design for disposal. If disposal is a problem, the product should be replaced by a more easily disposable product.

5. Use the least energy-intensive material that will do the job. Also, take into account the relative scarcity of materials.

6. Consideration should be given to the polluting effects of a material's manufacture.

Relative biological effectiveness (RBE). ◇IONIZING RADIATION, DOSE MEASUREMENT.

Rem. ◇IONIZING RADIATION, DOSE MEASUREMENT.

Reserves. The amount of substance, e.g. oil, that still remains available to be exploited. They are normally referred to as recoverable reserves, i.e. the total quantities of oil that are likely to be brought to the surface for commercial use within a certain time-span and level of technology. (◇RESERVES–PRODUCTION RATIO.)

Reserves–production ratio (R/P ratio). The ratio between the annual production rate and the proved recoverable reserves remaining in the ground, i.e. the number of years of production remaining at the rate of production in the year in question. Despite the impressive discoveries of oil, the R/P ratio has recently been falling steadily due to a continually increasing demand.

The R/P ratio which the oil industry likes to operate on is 20, to allow adequate time to find new fields and develop existing ones. Current discovery rates are 18 thousand million barrels per year and current production is 20 thousand million barrels per year. Hence, reserves are currently in a declining phase.

The current global R/P ratio for oil is 30,

which provides a buffer for a few years but unless prompt action is taken in developing alternative sources of supply the situation will turn sharply against us from the end of the decade. . . .

The last 25 years has seen the discovery of three quarters of the oil currently known in the Middle East and there is no other sedimentary basin remotely similar to the Middle East elsewhere in the world (Walters, 1974).

It should be noted that an R/P ratio of 20 does not allow much time to develop alternative energy sources and life-styles, or contain economic growth and stabilize populations.

P. I. Walters, 'Carbon and hydrogen sources – the supplier', *Chemistry and the Needs of Society*, Special Publication No. 26, The Chemical Society, 1974.

Reservoirs. The provision of storage capacity to remove fluctuations in the flow of a material or component.

Resources. Anything that is of use to man. (◊RESOURCES, RENEW-ABLE; RESOURCES, NON-RENEWABLE.)

Resources, Non-renewable. Substances which have been built up or evolved in a geological time-span and cannot be replaced except over a similar time-scale. Examples are copper, tin, COAL and OIL. It is often (erroneously) stated that when, for example, high-grade copper ore runs out, low-grade copper ore will become economically workable. However, this view neglects the facts of energy resources depletion and increasing pollution with lower grade burdens. Even if there were unlimited supplies of energy, the limitations imposed by the LAWS OF THERMODYNAMICS and climatic stability mean that there are limits to how much energy man may use in working low-grade ore deposits. Furthermore, ore sources do not necessarily become more plentiful with lower grades (◊ARITHMETIC–GEOMETRIC RATIO).

Recycling is one method of conserving finite resources. Some resources such as land, water and air have definite limits on the amount of exploitation they can sustain. Water is a renewable resource, but it is finite in its rate of supply as dictated by the hydrological cycle.

Resources, Renewable. Resources that derive from solar energy such as fish, trees, wind, rain. Plant-life such as timber or grass should be managed for MAXIMUM SUSTAINABLE YIELD. If the yield exceeds this rate, the system gives ever-diminishing returns. In a fishery pushed past its limit, the catch is maintained by collecting greater and greater numbers of younger fish until there is extinction. Resource management is now a necessity due to the pressures of POPULATION GROWTH and affluence.

Reverse osmosis. A means of separating solutions such as sugar in water or desalting by means of a suitable semi-permeable membrane and the application of pressure to the solution. (◊DESALINATION.)

Revolving screen. Separation device, used in gravel extraction, rock-crushing and domestic refuse RECYCLING, consisting of a perforated drum which allows materials of varying sizes to be graded and removed. ($\lozenge$ AIR CLASSIFIER.)

Right to discharge. An alternative to the principle that the polluter must pay is the proposal that any firm wishing to discharge a pollutant to the environment would have to pay for the right to discharge such pollutants. The interpretation is essentially a change in emphasis, from curative to preventive legislation, and would have the advantage of ensuring that firms maximize their benefits after payment. ($\lozenge$ POLLUTER-MUST-PAY PRINCIPLE; BEST PRACTICABLE MEANS.)

G. Nonhebel, 'Best practicable means and presumptive limits: British definitions', *Atmospheric Environment*, vol. 9, 1975, pp. 709–15.

Ringelmann charts. A series of six charts of graduated shades from white to black for assessing the darkness of a plume of smoke by visual comparison. The charts are a set of numbered grids differing from one another in the width and spacing of black lines printed on a white background. Ringelmann 1 is equivalent to 20 per cent black, 2 – 40 per cent black, 3 – 60 per cent black, 4 – 80 per cent black, and 5 – 100 per cent black. The charts must be used at a certain specified distance so that the grid forms a grey scale.

'Dark smoke' is smoke which is as dark or darker than Ringelmann 2, and 'black smoke' is as dark or darker than Ringelmann 4.

River regulation. Upland reservoirs in the upper reaches of the river are used to regulate the flow and especially to increase it at times of low run-off. Abstraction is carried out in the lower reaches close to the main areas of demand – the yield can be much greater than if water were taken from the reservoir directly. River regulation is a common means of augmenting water supplies.

Occasionally AQUIFERS are used as extra sources of water to augment the flow of a river in dry weather. The conjunctive use of aquifers and rivers has resulted in increased yields of water from the Thames and is also under survey for the Great Ouse. ($\lozenge$ WATER SUPPLY.)

Road traffic noise. The disturbing features of traffic noise are its general level and its variability with time. The latter refers to short-term variations due to the passage of individual vehicles, and to longer period variations at different times of day due to the general changes in traffic flow. It has been found that the dissatisfaction expressed by occupants of dwellings varies in accordance with the peak noise levels. Hence an index known as the '10 per cent level' (L_{10}) – log 10 – is used. L_{10} is the level of noise in dB(A) exceeded for just 10 per cent of the time. For the measurement and prediction of noise from traffic, the average of L_{10} values for each hour between 6 a.m. and 12 midnight on a normal weekday has been recommended in the UK as giving satisfactory correlation with dissatisfaction. This is known as L_{10} (18 hours), and permits accurate predictions for design purposes. Some values of L_{10} (18 hours) and typical conditions in which they are experienced are listed below.

L_{10} *(18 hour)* $dB(A)$	*Situation*
80	At 60 feet from the edge of a busy motorway carrying many heavy vehicles, average traffic speed 60 mph, intervening ground grassed.
70	At 60 feet from the edge of a busy main road through a residential area, average traffic speed 30 mph, intervening ground paved.
60	On a residential road parallel to a busy main road and screened by the houses from the main road traffic.

(◊NOISE; DECIBEL; HEARING; NOISE INDICES; AIRCRAFT NOISE; INDUSTRIAL NOISE MEASUREMENT.)

Rotary kiln. A slowly revolving drum, lined with refractory material and fired by gas or oil, used in the processing of ores and cement manufacture. Rotary kilns are also used to incinerate domestic refuse to obtain a high degree of 'burn-out'. A recent recycling innovation is to use domestic refuse as part of the fuel supply for cement manufacture in rotary kilns as the needs of refuse combustion and cement manufacture complement each other.

S

Safe yield (1).   In water supply, the long-term rate at which water can be extracted from an AQUIFER without a continuing progressive decline in its water level or other adverse effects.

Safe yield (2).   In fisheries, the allowable catch which will not endanger the breeding stock (◇MAXIMUM SUSTAINABLE YIELD).

Salinity.   The total amount of dissolved material expressed in terms of kilogrammes of material per million kilogrammes of feedwater, i.e. parts per million (ppm) of total dissolved solids.

Typical sea water has a salinity of 35000 ppm of which 30000 ppm is salt (NaCl). Now, accepted potable water standards are a maximum of 250 ppm as salt and a total dissolved solids content of 500 ppm and these figures should preferably be lower. Bearing this in mind, if a desalting process can sustain only a 90 per cent separation of total dissolved solids from a feed of 35000 ppm, then the product water will have a dissolved solids content of 3500 ppm. Water of this salinity is obviously not potable. This restriction applies to both electrodialysis and reverse osmosis but not to distillation, as product purity can range from 1 to 100 ppm from a feed of 35000 ppm (◇DESALINATION).

The salinity of sea and brackish waters is shown in the table below.

Source	Salinity Total dissolved solids (ppm)
Potable well water	300–500
Approximate limit for irrigation	1000
Brackish well water	1500–6000
Baltic Sea	2000–3000
Arabian Gulf	44000
Typical sea water	35000

Sanitary landfall.   ◇CONTROLLED TIPPING.

Schistosomiasis (Bilharzia).   The disease caused by the worms *Schistosoma haematobium* and *Schistosoma mansoni*, which use human beings as a primary host and snails as secondary hosts. Irrigation ditches make ideal transmission networks for the diffusion of these snails. To quote the World Health Organization:

The incidence of bilharzia has increased but it is of man's doing. As he constructs dams, irrigation ditches, etc., to alleviate the world's hunger, he sets up the ideal conditions . . . for the spread of the disease (WHO, 1961, p. 431).

(◇DAM PROJECTS.)

Scrap. Discarded material from manufacturing, processing or remnants after an article's useful life has run out. (◇RECYCLING; DOMESTIC REFUSE.)

Sea, Mining of. It is estimated that in the last 40 years we have used more of the earth's minerals than in the whole of previously recorded history. It is inevitable, therefore, that in their search for a continuing source of minerals the mining companies should turn their attention to the seas, which hold vast quantities of mineral wealth.

At present about a dozen minerals are extracted from the sea-bed on a commercial basis, including tin, iron, titanium, barium, calcium carbonate, bromine and magnesium. In addition, there is a major industry in sand and gravel recovery, with British coastal waters supplying not only this country's needs but also exporting about two million tonnes of material a year.

There is little doubt that the extraction of minerals from the sea is going to increase enormously over the next decade, one potential source being the mechanisms employed by many sea animals which have the capacity to concentrate minerals to a remarkable degree. These processes are little understood at present. For instance the proportion of copper in octopus ink is 100 000 times that in sea water; sponges can concentrate gold, and sea-squirts concentrate vanadium one million times. However the principal source of mineral wealth is undoubtedly the sea-bed. Diamonds can be dredged from the sea-bed at rates up to five times those obtainable on land, and manganese nodules are to be found in large quantities in certain areas.

Ignoring United Nations resolutions declaring the oceans' riches to be a common heritage for all, an American company, Deepsea Ventures Inc. of Delaware, laid claim to a 20 000 square mile piece of sea-bed between Hawaii and the Californian coast, in a small advertisement in *The Times*. A single square mile of ocean floor in the right place can be covered by up to 75 000 tonnes of manganese nodules, potato-sized lumps of material rich in manganese, copper, nickel and cobalt. At about £30 a tonne (a Deepsea Ventures' estimate) the piece of ocean floor the company has claimed could be worth up to £43 000 million. Deepsea Ventures plans to suck the nodules up from the sea-bed at the rate of one million tonnes a year and take them to a processing plant on the Gulf of Mexico. Others with similar ambitions include the late Mr Howard Hughes and a number of international mining companies.

If the plans do prove economic, there are dangers of ecological damage, and the serious risk that developing countries who now export the minerals from conventional mines will be squeezed out of business.

Another rich mineral strike has also occurred in the Red Sea, where a mineral-rich sludge is actually seeping from three major depressions, called 'hot-holes', which have been found in a rift in the sea-bed. An American company, Red-Sea Enterprises Inc., has claimed the mineral rights on 270 square miles of sea-bed at the bottom of the Red Sea. This one claim is estimated to be worth more than £900 million and is possibly 10 or 100 times bigger than this. The sludge found in this area is rich in gold, silver, copper and zinc and could represent the richest mining claim ever made.

The ecological disruption associated with sea-bed mining on the scale envisaged is at present difficult to estimate but is unlikely to be insignificant.

Sea water. ⬦SALINITY.

Second law of thermodynamics. ⬦LAWS OF THERMODYNAMICS.

Seed banks. ⬦GENETIC EROSION.

Selectivity (1). The characteristic of radiation, toxic chemicals or heavy metals (if ingested) to affect certain organs of the body to a much greater degree than the whole body dose would indicate. (⬦MINAMATA DISEASE; HALF-LIFE.)

Selectivity (2). The characteristic of plants and INSECTS to adapt to their environment by means of genetic selection of the most favourable strains.

Selenium (Se). A so-called heavy metal, atomic weight 78·96. Selenium is not, in fact, a metal but has certain metallic properties. It is a member of the sulphur group. It is produced as a by-product of the refining of copper, nickel, gold and silver ores. It is used extensively in the electronics industry, and also in paints and rubber compounds. It is a micro-nutrient at levels of 0·02 to 1 part per million. Maximum allowable concentration in drinking water is 0·01 part per million. It can act as a systemic poison. The LD_{50} (⬦LETHAL DOSE) for one selenium compound is as low as 4 microgrammes per kilogramme body weight. Known cases of poisoning are rare.

Sewage. The mixed liquid and solid effluent from domestic dwellings which is sent to sewers for treatment at sewage treatment plants. (⬦SEWAGE EFFLUENT TREATMENT.)

Sewage effluent treatment. The reduction of the organic loading (⬦BIOCHEMICAL OXYGEN DEMAND) that raw sewage would impose on discharge to streams and watercourses. It is carried out by oxidation of the sewage, i.e. contact with air, which oxidizes most of the wastes to allow discharge. Several steps are required:

1. Preliminary screening to remove large suspended solids, metal and rags.

2. Grit removal.

3. Sedimentation to allow as much suspended organic solids as possible to settle.

4. Biological oxidation in either of two plants:
   (a) percolating filter, i.e. a packed bed of clinker or stones 2 metres deep, through which the sewage trickles, and in which the surface area is maximized; or
   (b) activated sludge, in which the sewage is aerated in tanks by agitators which maintain the level of dissolved oxygen as high as possible. The sludge is recycled to seed the raw sewage and allow treatment to proceed faster.
   In both cases the aerobic bacteria are able to grow and the end result is sludge and an effluent very low in biochemical oxygen demand (20 milligrammes per litre or less). Discharge of the effluent, provided there is at least an eight-fold dilution in the river, should not normally cause any problems following step 5 below (and if necessary step 6 in cases where there are strict discharge requirements).

5. Sedimentation in chambers called humus tanks where the discharge from biological oxidation has most of its residual suspended solids removed. The discharged effluent is made to flow upwards through the tank at very low velocities (1 to 2 metres per hour) so allowing the suspended solids to be removed as sludge, which is then disposed of (⬦SLUDGE, SEWAGE for disposal processes). For the activated sludge process, a portion is recycled as shown in Figure 55.

6. Tertiary treatment. If the treated effluent is still of inadequate quality (e.g. too high in suspended solids) it can be microstrained in special filters. Tertiary treatment will grow in use as higher discharge standards

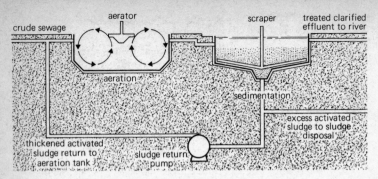

Figure 55. Activated sludge effluent treatment process.

are applied to reduce or contain the pollution of inland rivers from sewage treatment plant effluent. (◊ AEROBIC PROCESS.)

Sewage effluent treatment, Deepshaft process. This technique has been developed by I C I and is claimed to cut the costs of sewage treatment by up to 50 per cent. It relies on a deep shaft (approximately 135 metres) into which sewage and biodegradable effluents are admitted as shown in Figure 56. The mixture is circulated by air injection and the mixing processes so created enable the aerobic bacteria to reduce the biochemical oxygen demand of the wastes at rates of up to 10 times that of conventional sewage treatment plants. A baffle arrangement keeps untreated and treated effluent separate. The method still relies on aerobic bacteria but has increased the intensity of operation, thus allowing plants to be greatly reduced in size.

This plant is an offshoot of aerobic fermentation research where the tower or shaft is much shorter as the bacteria require less contact time with the substrate to be fermented. (◊ BIODEGRADATION.)

Silicosis. ◊ FIBROSIS.

Single cell protein (SCP). Protein manufactured by microbial means from organic substances such as oil, natural gas, cellulose or sewage. Microbial conversion offers the means of converting wastes into useful protein for both animals and man, because of the rapid rate of metabolism of micro-organisms and their subsequent growth and reproduction. Under suitable conditions many microbes, especially YEASTS, may be maintained in exponential growth situations in continuous culture with a doubling time of 15–20 minutes (◊ EXPONENTIAL CURVE).

The production technology uses aerobic FERMENTATION with copious supplies of air to obtain maximum growth rate. A single

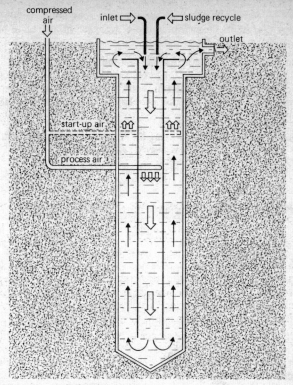

Figure 56. Deepshaft effluent treatment process (as developed by ICI).

strain of organism is chosen because this allows the optimum production of protein from the substrate, e.g. the yeast. *Candida utilis* is used for protein production from molasses and a similar strain is used on oil. The disadvantage of single strains is that sterility must be maintained.

SCP is a major growth area and is seen by many as the only route to alleviating shortages of feedstocks for both cattle and humans. (⬦SYMBA.)

Slag. The molten ash formed during smelting operations. It is also formed in high-temperature incineration processes where refuse is converted to a molten slag which is removed from the bottom of a special cupola. Steelworks slag is often processed and recycled because

231

of its high ferrous content. Other uses are road foundations and in sintering plants for the manufacture of lightweight aggregate building blocks.

Sludge, Sewage. The term for the residue after aerobic treatment of sewage has been completed (◊SEWAGE EFFLUENT TREATMENT). It is a slimy, offensive material with 95 per cent or greater water content. Its disposal is a major problem and four main methods are employed:

1. Conversion into METHANE gas by digestion, and subsequent use as a soil dressing (◊ANAEROBIC PROCESS).

2. Dewatering followed by incineration.

3. Spreading on land.

4. Dumping at sea from specially constructed sludge vessels.

Digestion is an ecologically sound method and is practised in large sewage works where the gas evolved is used to run the plant and heat the digesters. (Surplus gas may be available for sale.) The process renders the sludge free from pathogenic organisms and offensive odours and reduces by 30 per cent the weight of solid matter to be disposed of. Subsequent solids disposal – land spreading or dumping – is very much simplified. Incineration is costly and involves the use of fuel, as opposed to the generation of a fuel by the digestion process. It is often adopted where the sludge is produced in large volumes as in inland cities. Land spreading is sometimes used in areas where the sludge is relatively free from HEAVY METAL contamination. This may restrict its use on the same land to once in 15 years to prevent accumulation of toxic heavy metals in excessive quantities. Ocean dumping is used by large coastal cities; the sludge is dumped in deep trenches outside the tidal flow to be diluted by the sea.

Smog. Collective name for a FOG containing man-made air pollutants, usually trapped near the ground by a TEMPERATURE INVERSION. Constituents can be smoke, sulphur dioxide, unburnt hydrocarbons and nitrogen oxides. Smog should not be confused with PHOTO-CHEMICAL SMOG.

Smoke. Gas-borne solids resulting usually from incomplete combustion or chemical reaction. The particles are usually less than 2 microns in diameter. Other solid particles in smoke are compounds of silica fluoride, aluminium, lead, acids, bases, and organic compounds such as phenols.

The term 'dark smoke' is used to denote black dense smoke resulting from incomplete combustion. Smoke density is measured by photometric methods or estimated visually ($\diamond$ RINGELMANN CHARTS).

Smoke is synergistic with SULPHUR OXIDES and can have significant adverse health effects at concentrations as low as 200 microgrammes per cubic metre.

Smoke stain. This is a patch of dark material accumulating on a white filter paper after a measured volume of air has been drawn through it. The darkness or change in reflectivity of light from the white paper can be used as a measure of the amount of dark particulate material in the air.

Soil. The topmost layer of decomposed rock and organic matter which usually contains air, moisture and nutrients, and can therefore support life. Soil types include sand, clay, loam (a sand–clay mixture) and peat (which contains a large proportion of decaying plant matter). In tropical zones it may be lateritic, i.e. clay formed by the weathering of igneous rocks, which is notoriously difficult to cultivate – and once cultivated can set into a rock-like mass making further cultivation impossible. ($\diamond$ PLANT NUTRITION.)

Soil, Fertility and erosion of. The introduction of fertilizers and pesticides, combined with increased mechanization, has permitted the separation of arable and livestock farming. Arable land was no longer used for grazing, and livestock was increasingly housed indoors and fed concentrated rations.

Although this led to increases in productivity, by the 1960s there was evidence of a deterioration of arable soils, because of the use of heavy machinery used before the land was sufficiently dry, and a reduction in the amount of organic matter being returned to the soil. These effects, combined with the wholesale removal of hedgerows, have altered drainage patterns, so that field drains are required to avoid waterlogging.

The deteriorating capacity of such soils to retain moisture can result in the drying out of soils, so that recurrent cycles of flood and drought tend to occur on the lighter soils and wind erosion becomes a significant factor. Soil blows are now a problem in many farms in East Anglia and the east Midlands.

In the mid 1960s the National Farmers' Union conducted a survey, the results of which led to a full government inquiry whose results were published in January 1970 (*Modern Farming and the Soil*, Agricultural Advisory Council). It was concluded that a further increase in the intensiveness of arable farming would lead to further deterioration in soil structures and erosion, probably for only a small increase in yield.

The remedy is to increase the grass acreage and to reintegrate livestock and arable farming. ($\diamond$AGRICULTURAL ECONOMICS; MODERN FARMING METHODS.)

See M. Allaby, C. Blythe and C. Hines, *Losing Ground*, Friends of the Earth for Earth Resources Research, revised edn, 1975.

Soil moisture deficit (SMD). The difference between the moisture remaining in the soil at any time and the FIELD MOISTURE CAPACITY. The SMD is important in agriculture and determines whether irrigation is required or not. In the UK the mean annual SMD ranges from less than 12 millimetres (west coast) to more than 150 millimetres (East Anglia) where irrigation is now extensively practised.

Solar energy. The prospect of a low maintenance cost, low environmental impact, 'free' energy source is obviously enormously attractive, and a considerable amount of research into practical methods of converting the sun's rays directly into usable energy has been mounted in the last few years. Such techniques would have the further advantage of helping to redress the energy imbalance between the developed and underdeveloped countries, since a majority of the latter are situated in tropical zones.

Engineering studies of large centralized solar electrical power systems suggest that capital costs per unit of energy output are within striking distance of traditional sources of energy, assuming the usual improvements of technique that would inevitably be associated with a serious commitment to such a system. Such schemes, however, would be in many ways a misuse of this ultimate, and in many ways ideal, resource, the principal advantage of which is that it is freely available in sufficient intensity between latitudes $\pm 30°$. There would appear to be little point in designing vast centralized systems of energy conversion which then have to overcome distribution problems, when there is no technological reason why small individual units, capable of providing single dwellings with heating, lighting and cooking facilities could not be developed. Rudimentary forms of such devices are already in use in many countries and would almost certainly become competitive with conventional energy systems if properly developed on a commercial basis.

Such diffuse solar technology would obviously be ideal in developing countries where capital is not available for the development of large centralized power systems. Furthermore, such devices are not limited to tropical areas. Modern 'selective black' surfaces, which are highly absorbent through the visible spectrum but poor radiators in the far infra-red, can attain very high working temperatures even on a cloudy day and bring solar energy use to the northern hemisphere too.

Solar heating. A common means of putting the sun's energy to use is to cover a black water-filled metal panel with glass or plastic. The water is then heated by the GREENHOUSE EFFECT. Similar techniques apply to solar distillation where the sea or brackish water to be distilled is not enclosed in a metal panel and can therefore be evaporated. The glass sheet is cooler than the vapour evolved and therefore condensation takes place and the result is pure distilled water. Solar devices are essentially area intensive, because of the low solar radiation density.

Other methods of solar heating include the use of black-coated pool bases, thus allowing the body of water to warm up very cheaply. Reflectors, 'concentrating' lenses, etc., have been tried mainly to generate high temperatures in special situations. (◊SOLAR ENERGY.)

Solution mining. As the reserves of high-grade ores diminish, the prospects of recovering metals from low-grade ores and spoil heaps is receiving great attention. The means are varied but solution mining, whereby the derived metal is leached *in situ* and then recovered from the leachate, is one of the most promising (◊LEACHING).

This technique enables extremely low-grade ores to be mined. For example, copper ores with 0·5 per cent copper would require, by conventional means, extraction of the ore, crushing and milling, and at least 200 tonnes of rock would be removed per tonne of copper recovered. Solution mining would leach the copper from the rock *in situ* using a dilute sulphuric acid solution, after it had been fractured, say, by explosion (◊NUCLEAR MINING), or crushed into large lumps (not milled which is extremely expensive) and piled into mounds. The copper ion solution is then ponded where it is placed in contact with scrap iron and ION EXCHANGE takes place. The reaction is as follows:

$$Fe + Cu^{++} \longrightarrow Fe^{++} + Cu$$

iron copper ferrous copper
ions ions deposit
(in solution) (in solution)

The iron goes into solution and the copper is deposited on the bed of the pond. The pond is drained and the copper deposit removed. By this means low-value scrap iron is substituted for high-value copper. Now the process can stop there *but* a more useful technique is to regenerate the leach solution through the action of bacteria which rejoice in the name *Thiobacillus ferroxidans*, which can oxidize the ferrous sulphate to ferric (a form of iron) and release sulphuric acid in the process. The leach liquor is thus regenerated. This method is used in the extraction of uranium from low-grade ores in spoil heaps, old mine workings, etc.

The bacterial reoxidation process is the key to much of this work and the leaching techniques are now being applied to low-grade copper and nickel ores and the possibility of recovering aluminium from non-bauxite sources is also under consideration. It also has obvious environmental benefits where mining is done *in situ*.

This technique will grow as high-grade ore reserves decline. (◊ MATERIALS RESOURCES.)

Soot. Finely divided carbon particles which adhere together. Soot is often left in flues where fossil fuels are incompletely burned.

Sound. Periodic wave-like fluctuations of air pressure. The amount by which the pressure changes is known as the sound pressure (which is the 'mean' of pressure peaks generated by the waves), and the rate at which the fluctuations occur is the frequency. High frequency sound is characterized by screeches and whistles; low frequency sound by rumbles or booms.

The sound pressure level is a measure of the sound pressure differences using the logarithmic decibel (dB) scale. The nature of the decibel scale is illustrated in Figure 57. It will be seen that any tenfold increase of

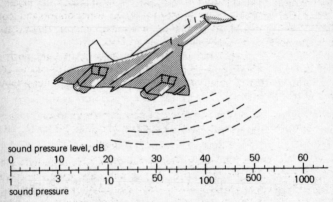

Figure 57. The decibel scale for sound pressures.

sound pressure on a linear scale corresponds to a rise of sound pressure level of 20 on the decibel scale. Thus, taking the pressure of a just-audible sound and ascribing to this a value of 0 dB, a sound of ten times that pressure has a level of 20 dB. A sound of a little more than three times the pressure of the just-audible one has a level of 10 dB. The sound pressure at which a sensation of pain begins in the ear is about one million times greater than that of the quietest sound that can be heard, and this has a level of 120 dB. Decibels therefore give a manageable way of measuring sound pressure, as well as having a close conformity to the ear's scale of response. The expression of a sound pressure level is always relative to the reference level of (in this instance) the quietest sound and is further discussed in the DECIBEL entry. (♢DECIBEL; HEARING; NOISE; ROAD TRAFFIC NOISE; AIRCRAFT NOISE; INDUSTRIAL NOISE MEASUREMENT; NOISE INDICES.)

Standard temperature and pressure (STP). As the density of gases depends on temperature and pressure, it is customary to define the pressure and temperature against which the volume of gases are measured. The normal reference point is standard temperature and pressure – 0°C at a standard atmosphere of 760 millimetres of mercury (approximately 100 000 pascal, the SI unit of pressure). All gas volumes are referred to these standard conditions. (♢GASES, PROPERTIES OF.)

Storage reservoir. A reservoir, normally constructed by damming a valley in an upland catchment area, for the provision of water for the public supply.

Stratification. The separation of a lake or sea into distinct layers or strata. These layers are characterized by 'warm' water on top and 'cold' on the bottom with an intermediate transitional band or 'thermocline' which is stagnant and seals off the bottom layer.

Figure 58 shows the stratification for Grasmere in the Lake District. Note that the amount of DISSOLVED OXYGEN has dropped to zero in the bottom layer and thus this anaerobic zone will not support aquatic aerobic life.

Stratosphere. The 'upper' portion of the earth's atmosphere above the TROPOSPHERE extending to a height of about 80 kilometres. The temperature is constant with height in this region.

Straw. Plant stems remaining after cereal crops have been harvested. In the UK the varieties are wheat, barley and oats; each one has its own characteristics. Barley and oat straw can be chopped, milled and fed to ruminants (sheep and cattle) as a carbohydrate source which requires protein supplementation for a balanced diet. As straw is produced in abundance in the UK (in 1972, 8 million acres produced 9 million tonnes), many uses have been proposed including papermaking, enzymatic decomposition to produce glucose and/or high protein feedstock. Currently, due to high animal feeding costs, it is likely that chemical delignification (♢LIGNIN) by boiling with caustic soda will become popular. The result is an easily digestible feedstock.

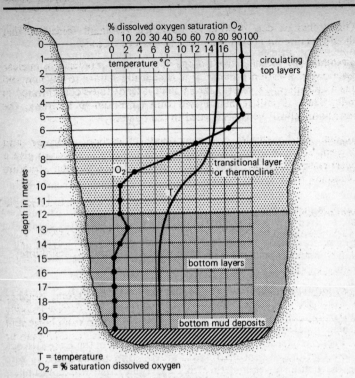

Figure 58. A stratified lake based on measurements of Grasmere, 2 September 1974. Temperature and dissolved oxygen are measured at metre intervals. (From *Land, Air and Sea – Research in Man's Environment,* N E R C, 1975.)

A microbiological method for breaking down straw for feedstuffs is to use selected strains of FUNGI. This technique can be carried out simply on the farm, so avoiding high transport and processing costs.

The process involves the following steps:

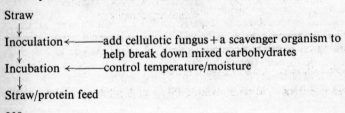

Straw
↓
Inoculation ← add cellulotic fungus + a scavenger organism to help break down mixed carbohydrates
↓
Incubation ← control temperature/moisture
↓
Straw/protein feed

This technique can provide an upgraded ruminant feedstock that should require no protein supplement. (◊CELLULOSE.)

Strontium–90. Radioactive isotope of strontium produced in nuclear reactions as a fission product. It is chemically similar to calcium and has a half-life of 28 years. Like calcium it tends to be concentrated in milk. Thus, if milk contaminated with strontium-90 is drunk, the strontium–90 will be concentrated in the bones.

Sugars. Carbohydrates, i.e. compounds of carbon, hydrogen and oxygen, which are usually crystalline and dissolve in water to give a sweet-tasting solution. They may be classified by molecular structure as mono-, di- or trisaccharides.

Monosaccharides
- hexoses (6 carbons)—$C_6H_{12}O_6$, e.g. glucose, fructose
- pentoses (5 carbons)—$C_5H_{10}O_5$, e.g. xylose

Disaccharides—$C_{12}H_{22}O_{11}$, e.g. sucrose, maltose, lactose, cellobiose

Trisaccharides—$C_{18}H_{32}O_{16}$, e.g. raffinose

Sulphate. Salts of sulphuric acid containing the SO_4 group. Sulphur dioxide emitted from the burning of sulphur in fuels is oxidized slowly in the atmosphere to sulphur trioxide (SO_3) which forms sulphuric acid with moisture and sulphates with basic materials such as ammonia or metals and their oxides. Sulphates so formed are particulates. Sea spray is a substantial natural source of airborne sulphate particles. The chemical reactions for acid formation (as well as acid attack on limestone) are as follows:

$$SO_2 + \tfrac{1}{2}O_2 \longrightarrow SO_3$$

Sulphur dioxide   Oxygen   Sulphur trioxide

$$SO_3 + H_2O \longrightarrow H_2SO_4$$

Water ⟶ Sulphuric acid

$$H_2SO_4 + CaCO_3 \longrightarrow CaSO_4 + CO_2 + H_2O$$

Sulphuric acid   limestone   Calcium sulphate   Carbon dioxide   Water

Hence buildings can be subjected to corrosive acid attack by the combustion of sulphur-containing fuels. (◊ACID RAIN.)

Sulphur dioxide. ◊SULPHUR OXIDES.

Sulphur oxides. Sulphur dioxide (SO_2) and sulphur trioxide (SO_3).

Of the two sulphur dioxide predominates and in the presence of particular catalysts conversion to sulphur trioxide can take place.

Sulphur oxides occur naturally from sources such as volcanoes, sulphur springs, and decaying organic matter. Annual global production by man is about 100 million tonnes per year with over 90 per cent produced in the northern hemisphere. Man-made sources are principally combustion of fuels which contain sulphur, brickworks and spontaneous combustion in coalmine spoil heaps. The man-made contribution is usually concentrated in industrial and domestic areas and can severely affect health during SMOG conditions. It is synergistic in combination with smoke. Together they affect the respiratory tracts and about 1 per cent of the population encounter bronchial spasms at concentrations of between 300 and 500 microgrammes per cubic metre (μg m^{-3}). Above 57 000 μg m^{-3}, waterlogging of the lungs takes place and eventually respiratory paralysis. During the 1952 London smog the deaths of 4000 people in one week and another 8000 in the following three months were attributed to the combination of sulphur dioxide and smoke. The maximum concentration of sulphur dioxide during the two-day period was 4900 μg m^{-3}, which is well within the threshold limiting value of 15 000 μg m^{-3}. The difficulties inherent in air pollution studies are highlighted by this example, as it was smoke and sulphur dioxide jointly that caused death (see Figure 59).

The effects of sulphur dioxide and smoke on man begin at a combined concentration of 300–500 μg m^{-3} for sulphur dioxide and 250 μg m^{-3} for smoke. But the effect of sulphur dioxide on LICHENS begins at 40 μg m^{-3}. Thus the damage to lichens does not correlate with the health effects on man. But where should the permissible level of exposure be set? The UK National Survey of smoke and sulphur dioxide shows that, in London and the Midlands, concentrations of sulphur dioxide occasionally exceed 500 μg m^{-3}, and in rural areas the background level often exceeds 40 μg m^{-3}, which may impair growth of crops.

It is clear that society puts up with different levels of effect, according to what is being affected – in this case plants or people – and it is apparent that sulphur dioxide levels are set for people. This may or may not be a wise judgement. The effects of lower sulphur dioxide levels may be increased crop yields, but the costs would be the costs of desulphurizing fuel before combustion, or the flue gases after combustion and before emission. COST–BENEFIT ANALYSIS could possibly be helpful in this area. This example illustrates how much is yet to be done to determine what is an acceptable level of any particular pollutant.

Removal of sulphur oxides from the atmosphere is usually accomplished in the form of ACID RAIN, with serious effects on property

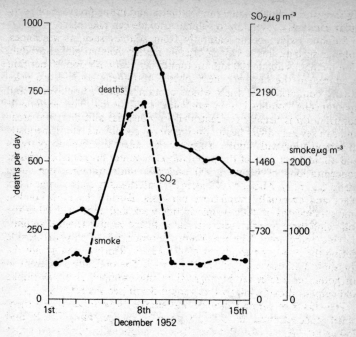

Figure 59. Health effects of SO₂ and smoke in December 1952 during London smog.

through corrosion and in some cases on aquatic ecosystems due to the alteration of pH. (◊SULPHATE.)

National Survey of Air Pollution 1961–72, Warren Spring Laboratory and Department of Trade and Industry, HMSO, 1972.

Sulphur trioxide. ◊SULPHUR OXIDES.

Supersonic flight. If the speed of an aeroplane is greater than the speed of sound in air then this is termed supersonic flight, and the plane is said to be travelling at greater than Mach 1 (the speed of sound in air). When this is the case, the plane creates a so-called shock wave and a noise or 'boom' accompanies the flight path. The boom intensity is a function of the speed, altitude and plane design. (◊CONCORDE.)

Supply curve. A supply curve shows, for a range of prices, the quantities of a commodity that will be offered for sale. Such a curve may be drawn for a single seller or for a number of sellers in a market. (⟡ DEMAND.)

Symba. A process, developed by the Swedish Sugar Company, which uses starch as a substrate for the eventual production of SINGLE CELL PROTEIN using the *Candida utilis* or *torula* yeast strain. For this purpose *Candida utilis* is propagated with an organism, Endomycopsin, which produces an ENZYME (amylase) which can split starch into sugars. The organisms grow in symbiosis, the amylase activity of Endomycopsin converting the starch to sugars which the *Candida utilis* – the faster grower of the two – uses as a fermentation substrate, i.e.

$$\text{Starch} \xrightarrow{\text{Endomycopsin}} \text{Glucose} \xrightarrow{Candida\ utilis} \begin{array}{l}\text{Yeast cell substance}\\(\text{single cell protein})\end{array}$$

The Symba process has been used in effluent treatment from potato processing plants and has reduced the BIOCHEMICAL OXYGEN DEMAND by 85 per cent in 10 hours. It is a means of reducing pollution and obtaining a useful product at the same time.

The process, or its counterparts, is under active investigation for single cell protein production from waste carbohydrates (cellulose, starch, 'simple' sugars), and from the organic substances in chemical industry waste streams. The waste fibres and bark from pulp and board production are also fruitful sources for production of single cell protein. (⟡ ENZYME TECHNOLOGY.)

Symbiosis. A compatible association between dissimilar organisms to their mutual advantage. The classic case is the association of nitrogen-fixing bacteria with plants of the clover family. The bacteria occupy nodules on the roots of the plants. The bacteria fix the nitrogen from the air into nitrates for the plants, who in their turn supply the bacteria with carbohydrates as an energy source.

Symbiosis can be put to commercial use as in the SYMBA yeast process. (⟡ SINGLE CELL PROTEIN.)

Synergism. A state in which the combined effect of two or more substances is greater than the sum of the separate effects, e.g. smoke and sulphur dioxide. It is the opposite of ANTAGONISM. (⟡ PARTICULATES; SULPHUR OXIDES.)

T

Tar sands. The Alberta tar sands (oil sands) constitute the largest known reserve of petroleum in the world (approximately 900 000 million barrels of 'in place' heavy oil). Total world resources are approximately one-fifth to one-seventh of the world's resources of liquid petroleum, but in general are considerably more costly to extract. Recent increases in oil costs, however, will probably ensure that efforts will be made to improve existing processing technology and techniques for extracting and upgrading known deposits.

Essentially the extraction process is one of open-cast mining, in which the 'overburden' of vegetation and earth must first be removed in order to expose the bitumen-soaked sand deposits. These deposits are extremely difficult to handle, being sticky and corrosive in summer and rock-hard in winter. The capital investment required for the large-scale extraction of such material is huge, and present techniques are probably only capable of recovering about 10 per cent of the synthetic crude which is potentially available in such deposits. (◊ OIL SHALES.)

Temperature. A property of a body or substance and a measure of how 'hot' it is, or how much thermal energy it contains. Temperature is measured on several scales: e.g. the centigrade or Celsius and Fahrenheit scales are both measured from a reference point – the freezing point of water – which is taken as 0°C or 32°F. The boiling point of water is taken as 100°C or 212°F respectively. For thermodynamic devices, it is usual to work in terms of absolute or thermodynamic temperature where the reference point is absolute zero, which is the lowest possible temperature attainable. For absolute temperature measurement the thermo-dynamic or kelvin scale which uses centigrade divisions is used. (Degrees Rankine, which uses Fahrenheit divisions, is now virtually obsolete in the UK since the adoption of SI.) To convert from centigrade to kelvin, add 273 to the centigrade reading. By convention no degree symbol precedes the abbreviation K.

Temperature is a fundamental measurement in most pollution work. The temperature of a stack gas plume, for example, determines its buoyancy and how far the plume of effluent will rise before attaining the temperature of its surroundings. This in turn determines how much it will be diluted before traces of the pollutant reach ground level.

Temperature inversion. The temperature of the air normally decreases at increasing heights above the ground. A variety of meteorological conditions can occur to reverse this trend and cause a layer of warmer air to overlie a cooler layer. The cooler air cannot then rise because it is heavier and so any air pollutants emitted below the inversion layer are trapped. Inversion of the normal temperature profile can begin at any height above the ground but the lowest are more noticeable in their effect as they trap the smoke and fumes from domestic chimneys. When they occur between 150 and 900 metres above the ground, they can trap the discharges from all chimneys except those of the largest power stations.

Inversions commonly occur in valleys or basins due to radiation cooling of the ground at night. This cools the air near the ground, and being cooler, it can drain into the low-lying areas as a katabatic flow.

Figure 60 shows the temperature profile with height for typical no-inversion and inversion conditions. In extreme circumstances a halt may be required on industrial processes until the inversion lifts. In practice this temperature profile can assume a wide variety of forms.

A related phenomenon associated with industrialized cities is the heat island. This is most noticeable at night and early mornings and is caused by hot air layers forming at building or chimney height level which is warmer than both ground conditions and the air above the layer. This heat island can be 5° to 7°C warmer than ground conditions and, of course, can trap any pollutants emitted within it. The heat islands are dome shaped and can disperse towards midday when the temperature increases, but in turn they may be replaced by a higher level inversion.

Teratogen. An agent that causes birth defects. A strong teratogen is DIOXIN found in herbicides. Guinea pigs fed with dioxin have a 50 per cent chance of survival at concentrations as low as 600 parts per million million. Fish in Vietnamese waters (1970–71) had concentrations on average of 540 parts per million million as a result of the use of herbicides as defoliants (1962–70). The use of defoliants in that country has been suggested as being responsible for rises in stillbirths and birth defects.

Tetraethyl lead (TEL). Tetraethyl lead and tetramethyl lead (TML)

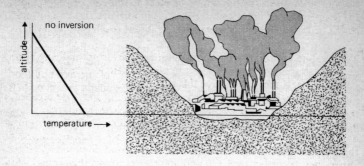

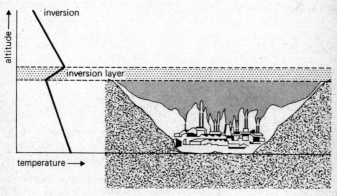

Figure 60. The effects of a temperature inversion.

are the principal additives to petrol to raise the octane rating and thus reduce the tendency to 'knock' in spark-ignition internal combustion engines.

Therm. A measure of heat (thermal energy) and the common basis for charging for heat supplied by gas or steam in the UK. It is 100 000 (10^5) British Thermal Units or 1.055×10^8 joules. It will eventually be superseded by megajoules (MJ) when SI is established.

Thermal pollution. The heat released from the combustion of fossil fuels or the dissipation of energy from prime movers, such as electric

245

motors, which is eventually converted to heat. All such releases end up as waste heat in the sinks of air and water.

Direct thermal pollution of water usually occurs at power stations where 60 per cent or more of the heat content of a fuel ends up as waste heat which must be removed by cooling water which is then discharged to rivers or coastal waters. This form of thermal pollution depletes DISSOLVED OXYGEN and can change aquatic ecosystems. It also increases the BIOCHEMICAL OXYGEN DEMAND and therefore *reduces* the capacity of a stream to assimilate organic wastes. The waste heat from power stations could be used for glasshouse or DISTRICT HEATING, but the heating load and power station capacity and duty require careful matching.

Thermal pollution of the atmosphere can cause local instabilities and in North America there is reason to believe that the industrialized Atlantic seaboard has its own man-made climate induced by the energy released. This changed atmosphere may initiate large-scale atmospheric motion between land and sea if regional changes such as those detected in America continue. Already the frequency of rain-storms near the steel-making town of Gary, Indiana, has increased considerably.

Thermal processing. A collective term for the disposal or conversion of domestic refuse or wastes by INCINERATION or PYROLYSIS.

Thermal wheel. A slowly rotating brick-lined heat exchanger which recovers heat from (say) furnace exhaust gases on one half, while the other half is giving up its recovered heat to incoming gases such as preheating air for combustion in boilers or blast furnaces. Recovery of 30–50 per cent of the heat in exhaust gases can be obtained.

Thermionic converter. A device for DIRECT ENERGY CONVERSION which gives an electrical current from the electrons emitted by the heating of a suitable metal. This thermionic emission produces a net flow of electrons from a high temperature electrode (the cathode) to the lower temperature electrode (the anode), both being enclosed in an evacuated enclosure.

However, the efficiency of the device is low unless the gap between the electrodes is very small (less than 25 microns). In addition, as radiation takes place from the hot surface to the cold surface, large temperature differences cannot be used.

Thermodynamic temperature scale. ◊ TEMPERATURE.

Thermoelectric converter. A device for DIRECT ENERGY CONVER-SION which gives an electric current when the two junctions of a loop of two wires of different metals are kept at different temperatures. This is the principle of the thermocouple for temperature measurement. However, for ordinary metals the voltage produced per unit of temperature difference is extremely low and it is only with the advent of semiconductors that the voltage obtainable rose from microvolts per degree kelvin to millivolts per degree kelvin (a factor of 1000). Semiconductor technology also allows vast numbers to be connected in series, so that useful outputs can be obtained. The electrical charge carriers can be electrons as in metals (or in semiconductor terminology 'n-type' materials) or they can be positive (or 'p-type' materials). Thus an 'n–p pair' with low resistance to heat flow and low electrical resistance would make the ideal unit for thermoelectric converter construction. The semiconductor is indeed such a device and just as for the THERMIONIC CONVERTER the 'working fluid' is the flow of charge carriers, so that the Carnot restrictions also apply ($\lozenge$CARNOT EFFICIENCY).

The thermoelectric device operates at lower temperatures than the thermionic converter and is therefore suited for waste-heat recovery applications as in, say, the exhaust gas stream of a gas turbine. Practical efficiencies are 10 per cent or less due to the temperature restrictions but this can be a useful addition in a total energy situation. The USSR manufactures a thermoelectric generator powered by a paraffin heater for radio receivers in rural areas.

Third World. $\lozenge$UNDERDEVELOPED COUNTRIES.

Three minute mean concentration. The maximum permissible concentration of pollutant at ground level from a stationary source, i.e. the concentration of pollutant over three minutes under the worst envisaged atmospheric conditions. For example, sulphur dioxide from a factory chimney is often specified as a three minute mean value. Yearly averages are likely to be between 5 and 15 per cent of this value, and monthly averages up to 25 per cent owing to wind changes and plume-dispersal effects.

The THRESHOLD LIMITING VALUE is often used to specify the maximum three-minute mean concentration which is usually set at TLV/40 or TLV/30. The values at which the pollutant concentration is measured are important, i.e. pressure, temperature and carbon dioxide values are required. For example, a pollutant measured as 1 milligramme per cubic metre ($mg\,m^{-3}$) at 300°C and 5 per cent carbon dioxide in the gas (e.g. flue gas inlet condition) has a concentration of 5·6 $mg\,m^{-3}$ at STANDARD TEMPERATURE AND PRESSURE and 10 per cent carbon dioxide. Thus, both ground level and stack gas measurements should be made.

Three minute mean concentration (TLV/40) values for five main gaseous pollutants (μg m⁻³)

Sulphur dioxide (SO_2)	325
Nitric oxide (NO)	750
Nitrogen dioxide (NO_2)	225
Carbon monoxide (CO)	1375
Hydrochloric acid (HCl – gas or fume)	175

Three minute mean concentration (TLV/40) values for particular pollutants (μg m⁻³)

Antimony (Sb)	12·5	Manganese (Mn)	125·0
Arsenic (As)	12·5	Mercury (Hg)	1·25
Beryllium (Be)	0·05	Molybdenum (Mo)	125·0
Cadmium (Cd)	2·5	Nickel (Ni)	25·0
Chromium (Cr)	25·0	Selenium (Se)	5·0
Cobalt (Co)	2·5	Tin (Sn)	50·0
Copper (Cu)	2·5	Vanadium (V)	1·25
Iron (Fe)	250·0	Zinc (Zn)	125·0
Lead (Pb)	3·75		

Note the very low limits on beryllium and the HEAVY METALS.

In addition to the three minute mean concentration or TLV/40 value, an absolute mass emission limit may also be placed on the total amount of a pollutant that may be emitted from a stationary source per hour. Thus, the mass emission limit from a smelter discharging less than 85 cubic metres per minute of gases up the stack is set at 45 kilogrammes of lead per 168 hours which may not approach anywhere near the TLV/40 value, but is required to prevent despoliation of land through build-up of lead with its dangers to animals. (◊EMISSION STANDARDS.)

Threshold. In environmental terms the dividing line between exposure to, say, a noise level above which physical damage may take place and below which damage will probably not take place. The threshold is a concept that is applied in many areas of industrial exposure to airborne pollutants (◊THRESHOLD LIMITING VALUE). However, the basis is that below the threshold none but the most sensitive or infirm would suffer. There are substances for which many believe there is no threshold effect whatsoever, e.g. VINYL CHLORIDE and ASBESTOS. IONIZING RADIATION is another area where a threshold may in effect be postulated by a recommended dose limit but the only safe exposure is no exposure at all. (◊CARCINOGENS; MUTAGENS.)

Threshold limiting value (TLV). In the UK, the threshold limiting value of an airborne pollutant is the maximum concentration of that pollutant (or mixture of pollutants) to which it is believed 'healthy' workers in industry may be repeatedly exposed day after day without adverse health effects, on an eight-hour day. For example, the threshold limiting value for CARBON MONOXIDE is 30 parts per million – a value which is often exceeded in peak-hour traffic flows in city streets.

The TLV is measured in the UK by the Factory Inspectorate and compared with American values which are revised annually. The TLV is not set for the public at large. In this case the THREE MINUTE MEAN CONCENTRATION is specified as a concentration at ground level which should not be exceeded in predicting the maximum ground-level concentration from some air pollution source. The three minute mean is often (but not always) expressed as a fraction (usually one-thirtieth or one-fortieth) of the TLV.

Tidal power.   Generation of power by the ebb and flow of the tides. This is an example of an 'income' source of energy and takes two forms:

1. Barrages with conventional hydroelectric turbines installed in special sluices, so that the incoming and outgoing tides generate power. The availability of suitable tidal locations is severely restricted and, as with WATER POWER, this form is not expected to grow significantly.

2. Floating power stations which rise and fall with the tide and have wave-powered floats or paddles which can drive water turbines for power generation. This method is not restricted to the availability of sites as for barrages, and, though still in embryonic form, may have some potential in the long term.

Time lag.   The time needed for an event or disruption to manifest itself. There are many examples in environmental matters of a time-lag effect. One such example is the use of nitrogen fertilizers on agricultural land from which water filters into the London Chalk Basin. For many years water from the aquifer in this area has been within World Health Organization limits for nitrates; now after 20 years or more of fertilizer application some wells are showing higher than normal nitrate levels.

The latency period for the development of cancer is well known. This can take tens of years, dating from exposure to the carcinogenic agent in the first instant. It is for this reason that before the introduction of new health products, food additives, hair dyes, etc., the most exhaustive tests must be carried out for mutagenic and carcinogenic properties. Hair dyes are a good case in point; they have been implicated as being carcinogenic but confirmation will have to wait for at least a decade in all probability.

The time-lag effect often obscures the cause of the problem encountered and leads to treatment of the symptoms only. (◊CARCINOGEN; MUTAGEN.)

Total energy. The integrated use of all or most of the heat involved in the combustion of fossil or nuclear fuels. Thus, instead of just generating electricity, a total energy scheme would generate electricity and sell heat for both DISTRICT HEATING and factory processes.

Toxic action of pollutants. There are three main mechanisms by which the human organism is affected by toxic pollutants.

1. They influence enzymatic action by, for example, combining with the enzyme so that it cannot function (◊ENZYMES IN THE HUMAN ORGANISM).

2. They can combine chemically with the constituents of cells, as, for example, carbon monoxide combining with blood haemoglobin so that oxygen transport to the brain is affected.

3. Secondary action because of their presence. Hay fever is brought about by pollen and the system reacts to produce histamine.

The factors of importance are the concentration of the pollutant, the length of exposure, the age, the activity – whether slight or heavy exertions – and the health of exposed person/population.

G. L. Waldbott, *Health Effects of Environments Pollutants*, C. V. Mosby, St Louis, 1973.

Toxic wastes. Those wastes usually emanating from industry such as cyanide compounds, HEAVY METALS, CHLORINATED HYDRO-CARBONS, which if dumped or disposed of indiscriminately can, for example, jeopardize water supplies by infiltration and be a public health threat.

Following 'fly tipping', i.e. unauthorized tipping of cyanide compounds on children's playgrounds and roadsides in UK in early 1970s, stringent procedures are now enforced so that the producers of toxic wastes dispose of them in an authorized manner by authorized contractors on approved sites.

The complexity of and dangers from toxic wastes from manufacturing processes is indicated by the accident which occurred at the Pitsea, Essex, toxic waste tip in early 1975. A load delivered earlier to the tip contained sodium hydrosulphide, sodium sulphide, sodium thiosulphate, sodium hydroxide, and traces of water and organic materials. Fifteen minutes later another load was tipped in the same spot; it contained a solution of aluminium oxide in sulphuric acid, with traces of lead, nickel, copper and titanium. The result was a chemical reaction which gave off hydrogen sulphide gas, which causes acute asphyxia,

and resulted in the death of a lorry driver ($\Diamond$ASPHYXIATING POLLU-
TANTS).

The need for sites such as Pitsea is obvious, the pressures on it are immense. In 1974, 70 million gallons of mainly toxic wastes were dumped there, a reflection both on the lack of alternative disposal facilities and the types of waste products from UK industries.

Toxic waste disposal is mainly by tipping in approved locations, such as Pitsea which has an impervious clay base and therefore groundwater pollution should not occur. Other methods are incineration or ocean dumping – which is to be deprecated as there is no guarantee that marine food chains will not be affected.

Trace elements.   Elements which occur in minute quantities as natural constituents of living organisms and tissues. They are necessary for the maintenance of growth and development; the shortage of any one may result in reduced growth, physiological troubles and eventually death. However, in large quantities they are generally harmful.

Trace elements include lead, silver, cobalt, iron, zinc, nickel and manganese.

Transferable drug resistance. $\Diamond$ANTIBIOTICS.

Transpiration.   Water is transferred from the soil to the leaves of plants by capillary action and osmosis. At the leaf surface the water transpires or evaporates and the vapour diffuses to the atmosphere. As evaporation also takes place from the surfaces of lakes and rivers, it is common to use the term evapo-transpiration to account for the land-based water vapour component of the HYDROLOGICAL CYCLE.

Trichlorofluormethane (CCl_3F).   A member of the group known as halogenated fluorocarbons, used as an AEROSOL PROPELLANT. Between 10 000 and 100 000 tonnes are liberated per year. It is assumed to break down chemically, releasing free chlorine which is said to be combining with ozone to deplete the earth's OZONE SHIELD. However, due to TIME LAG, the effects if any may not be evident for more than a decade.

Trichoderma viride.   A fungus which is capable of breaking down crystalline cellulose by the production of an ENZYME of the cellulase group which can accomplish the decomposition in a few days. Mutant strains of *Trichoderma viride* have been developed which accelerate the decomposition. *Trichoderma viride* is the basis of the Natick process for the production of glucose from cellulose ($\Diamond$ENZYME TECHNO-LOGY).

Trippage.   The number of trips that a returnable bottle makes in its lifetime. In the UK, average trippage is 25 times for a milk bottle; in the USA, 10 times for a soft drink bottle. ($\diamond$RECYCLING.)

Trophic levels. $\diamond$FOOD CHAIN.

Troposphere.   The 'lower' portion of the atmosphere about 8 kilometres high at the poles and 16 kilometres high at the equator. In the troposphere the earth's temperature decreases with height. ($\diamond$STRATOSPHERE.)

Typhoid.   Water-borne disease caught by drinking sewage-contaminated water. ($\diamond$PUBLIC HEALTH.)

U

Unburnt hydrocarbons. Airborne particles of hydrocarbon fuels not consumed in combustion and which are emitted in exhaust or flue gases particularly from INTERNAL COMBUSTION ENGINES. Main components are carbon monoxide, paraffins, olefins, and aromatic compounds. (⟡AUTOMOBILE EMISSIONS; PHOTOCHEMICAL SMOG.)

Underdeveloped countries. The categories 'developed', 'developing' and 'underdeveloped' are arbitrary divisions and tend to hide the great differences that may exist from nation to nation and between groups within nations. There is in fact a gradual transition from the poverty of Bangladesh to the affluence of North America. The following approximate criteria are probably as appropriate as any in helping to categorize the degree of development of a given country.

Underdeveloped nations
Per capita income less than £300 per head per annum.
More than half the population engaged in agriculture.
Diet usually inadequate.
More than half the population illiterate.
Life-expectancy not much more than 40 years.
Infant mortality between 100 and 200 per 1000 births.
High proportion of population under 15 years old (⟡AGE-SPECIFIC BIRTH AND DEATH RATES).
Exports: raw materials and protein.
Relatively large rate of increase in population (up to 4 per cent per annum).

Developing nations
A more difficult category to define. It implies the (arrogant) assumption that such nations actively wish to 'develop' the type of society with which we are familiar. Thus China, although sometimes categorized as a 'developing' nation, is unlikely to aspire to the 'affluent society' of the West. The Chinese (one devoutly hopes) probably consider themselves as 'developed', in terms of material acquisitions, as they want to

be. Nations traditionally referred to as 'developing' (e.g. Turkey, Brazil, Spain), as far as the standard criteria given above and below are concerned, come somewhere in between the underdeveloped and developed nations. Whether, as the resource crisis deepens, they will continue to develop probably depends on whether they have a scarce resource to bargain with (e.g. Middle East oil, Chilean copper, Moroccan phosphates).

Developed nations
Per capita income greater than £800 per annum.
Less than 10 per cent of the population employed in agriculture.
More than adequate diets with significant obesity problem.
Less than 10 per cent of the population illiterate.
Life expectancy around 70 years.
Infant mortality around 20 per 1000 births.
More even distribution of age groups.
Imports: raw materials and protein; exports: artifacts and technological skills.
Population growth small or zero.

It is highly improbable that any of the underdeveloped nations will ever acquire an adequate standard of living while their rapid population growth continues. Only China appears to have found the solution to population control – by means which, while effective, may be too draconian for most other countries. (◊ POPULATION GROWTH.)

United States of America, Energy requirements. ◊ 'PROJECT INDE-PENDENCE'.

V

Vanadium (V). Hard, white metal, used as a steel alloying element and as a chemical industry catalyst. The threshold limiting value is 500 microgrammes per cubic metre for dusts and 50 microgrammes per cubic metre for fumes of vanadium pentoxide (V_2O_5).

Vanadium affects most metabolic processes in the human organism. The lethal dose is between 60 and 120 milligrammes. Chronic exposure to environmental air concentrations of vanadium can lead to bronchitis.

Vector. An organism (animal, fungus) which transmits or acts as a carrier of parasites, e.g. the Anopheline mosquito is the vector for the malaria parasite, the Aedes mosquito the vector for yellow fever, the rat flea the vector for plague, and the tsetse fly the vector for sleeping sickness.

The use of insecticides to eliminate such vectors, together with generally improved standards of medical practice, have resulted in significant reductions in death rates of the UNDERDEVELOPED COUNTRIES. Unfortunately birth control schemes initiated in the same period have been far less effective and as a result populations in these countries are now increasing at up to 4 per cent per annum. (⟡ POPU-LATION GROWTH.)

Vinyl chloride (CH_2CHCl). A colourless gas used in the manufacture of polyvinyl chloride (PVC), a PLASTIC. The vapour has been found to be highly carcinogenic and has been linked with the deaths of 20 workers from angiosarcoma, a rare form of liver cancer, on a global basis. In the UK there is evidence that the vapour has caused impotence, stiffening of the joints, bad circulation and shortness of breath. In the USA very strong standards set exposures at a maxiumum of 5 parts per million and time-weighted average exposures as low as 1 part per million. This has also been adopted by Sweden. In the UK the threshold limiting value has been set at 10 parts per million (4 milligrammes per cubic metre), considerably higher than the US limit.

There is a great concern at the potential health effects of this substance and no threshold effect is postulated for it as for other pollutants, i.e. exposure to any concentration may cause damage. (◇MUTA-GEN; CARCINOGEN.)

Vitrification.   ◇NUCLEAR WASTES.

Volatilization.   The evolution of vapour (or gas) due to heating of a liquid or solid. A substance of high volatility (e.g. petrol) vaporizes easily and, in the case of inflammable liquids, is characterized by the FLASH POINT.

W

Waste factor. A term used in ENERGY ANALYSIS to measure the departure from ideal energy needed to effect a transformation. It is defined as

$$\frac{\text{actual energy required to effect transformation} - \text{ideal energy} \ldots}{\text{actual energy} \ldots}$$

Ideal energy is obtained from physical or chemical calculations which show the theoretical energy consumption required to manufacture a given component or product.

A waste factor close to 1 represents a very wasteful and inefficient process. Thus, it can be used to show scope for improvement in energy utilization.

Product	Actual energy (1968) MJ/Kg	Ideal energy	Waste factor
Iron	25	6	0·76
Petrol	4·2	0·4	0·90
Paper	38	0·2	0·99
Aluminium	190	25	0·87
Cement	7·8	0·8	0·90

From E. Gyftopoulos *et al.*, *Free Energy Use, Actual and Ideal*, Thermo-Electron Corp., Waltham, Mass., 1974.

Waste heat. ◊ THERMAL POLLUTION; ENERGY, EFFECT OF CONVERSION ON CLIMATE; DISTRICT HEATING.

Wasteplex. A concept of a highly integrated centre for the reception of domestic and industrial wastes from population catchment areas of around 500 000 people. This allows economic sorting of the wastes into directly and indirectly reusable fractions. (◊ RECYCLING.)

Water. One of the prime resources for life which must be preserved from contamination for public WATER SUPPLIES. It has many industrial uses: a conveying medium for wastes, slurries, wood-pulp, etc.; a heat-exchange medium; it is used in steam raising and also as a solvent. Its basic chemical formula is H_2O, yet it has other molecular combinations such as H_8O_4, which gives it many unique properties such as expansion on freezing.

For the purpose of this entry the chemical, physical and biological attributes are considered.

1. Chemical properties have a great influence on the use for drinking or industrial purposes or the support of aquatic life. The main parameters are:

(i) pH – a measure of the acidity or alkalinity ($\diamondsuit$ pH). Natural water supplies are usually in the pH range 6–8, whereas industrially contaminated waters can have any value in the range pH 1–13 and require neutralization to pH 7 before discharge.

(ii) Hardness – this prevents soap lathering properly. Also when the water is heated, scale may be deposited on the heating surfaces, which can ruin the efficiency of heat transfer and in extreme instances cause burn out of boiler tubes or heat exchangers. The hardness is measured by the concentration of calcium ions (Ca^{++}) and magnesium ions (Mg^{++}), present usually as calcium carbonate ($CaCO_3$) and magnesium sulphate ($MgSO_4$).

(iii) DISSOLVED OXYGEN and, related to this, the oxygen demand. These values are of great importance to biological systems ($\diamondsuit$ BIO-CHEMICAL OXYGEN DEMAND).

2. Physical characteristics. The colour, taste and appearance of water are of great importance in its suitability for consumption. The solids in solution and in suspension play an important part. The following characteristics are measured:

(i) Turbidity – an indication of the presence of colloidal ('gluey' suspensions of fine particles) particles such as silt or bacteria from, say, sewage treatment.

(ii) Colour – usually due to dissolved substances such as peat acids which give highland streams their brown appearance.

(iii) Taste and odour due to the presence of dissolved solids or gases, e.g. 0·01 milligrammes per litre (1 part in 10 million) of phenolic liquors can taint water and render it unpalatable. The gases can be of biological origin from algae or from bacterial action on wastes or impurities.

(iv) Temperature – this has a direct bearing on the saturation concentration of dissolved oxygen in the water; the greater the temperature, the less oxygen the water can hold in solution and therefore temperature has important biological consequences.

3. Biological properties. The variety and type of water organisms and aquatic life give a very good indication of its biological state of health and the BIOTIC INDEX is used as such a measure. Other tests are made to indicate the presence of various classes of micro-organism algae and bacteria, and especially the bacteria, *Escherichia coli*, which indicates the presence of faecal matter (◊COLIFORM COUNT).
(◊WATER SUPPLIES; WATER TREATMENT; PUBLIC HEALTH.)

Water consumption. The quantity of water abstracted and used for any purpose, irrespective of the state or time in which it is returned to the source or to the atmosphere. Per capita consumption is the amount of water supplied to an area divided by the size of the resident population. Some indication of the daily quantities of water abstracted in England and Wales for 1971 (Water Resources Board) is given below.

	Million cubic metres per day
Public water supplies	14·3
Electricity generation	18·9
Industry (two-thirds from public supplies)	9·2
	42·4

Abstraction for electricity generation for cooling purposes is mainly from rivers.

Eighth Annual Report for the Year Ending 30 September 1971, Water Resources Board, HMSO, 1972.

Water power. The free renewable source of energy provided by falling water has been put to use for centuries. Hydroelectricity is practised in many countries, yet in the USA there are only five plants with capacities in excess of 1000 megawatts. In the UK it supplies less than 2 per cent of the nation's power needs and there is little scope for further development as almost all sites have been used. It is capital intensive, requiring costly dams and earthworks which may have a finite lifetime due to silting up behind the dam. This will restrict its use at any one site to 100–200 years unless silt-laden floodwater is diverted to lengthen the life of the reservoir. Globally there are locations that can, in total, supply energy comparable to the present rate of energy consumption. South America, Africa, Asia and the USSR account for 80 per cent of the potential capacity. Because of the high capital investment per megawatt capacity, its contribution to future energy supplies is not expected to be significant in the West.

259

Water quality standards. Water for domestic use should be free from pathogenic bacteria and other disease-carrying organisms ($\lozenge$COLIFORM COUNT) and also comply with purity standards similar to those in the following table.

World Health Organization water purity limits

Substance	Maximum concentration permissible in public water supplies – parts per million
Carbon dioxide	20
Carbonates of sodium and potassium	150
Chlorides	250
Chlorine (free)	1·0
Copper	3·0
Detergents	1·0
Fluorine (as fluorides)	1·5
Iron	0·3
Lead	0·1
Magnesium	125
Nitrates	10
Phenols	0·001
Sulphates	250
Zinc	15
Total solids in suspension	500
Maximum NaCl	250

The World Health Organization quality standards are often infringed in arid zones where water supplies may contain in excess of 1000 parts per million (ppm) total dissolved solids. For most supply purposes 500 ppm is taken as the maximum permissible upper limit. It is useful to put the upper limit of 500 ppm in perspective. A typical stream in the UK can have 200–300 ppm total dissolved solids prior to abstraction. After passing through the human organism and household use and then to sewage treatment plants, the sewage plant effluent can have another 200 ppm dissolved solids added due to salt in the human diet, the breakdown of organic matter in plants, etc. Thus for a *hypothetical* case where a river has an initial total dissolved solid content of 200 ppm and an abstraction rate of 50 per cent of the total flow is practised, on the return of the effluent stream to the parent, the total dissolved solids content will rise to 300 ppm for a once through use. If a river serves several conurbations downstream each practising abstraction (at say 50 per cent of the flow rate) and adding an effluent with an additional 200 ppm, then only three abstraction cycles are possible before the

500 ppm dissolved solid limit is met. This *very simple* example shows the inevitable degradation which takes place during use. Demineralizing processes may be necessary before repeated (or any) abstraction can be practised on some UK streams. The Trent is a good example: it receives sewage effluent from Birmingham plus industrial discharges. The net result is that the Trent at Nottingham is unfit to drink and a resource which could yield up to a million cubic metres per day remains currently unavailable. Demineralizing processes are being tested for suitability on the Trent water. Thus even in Britain water can be a resource which may require DESALINATION techniques similar to the arid and semi-arid zones. The Colorado river is another example of the degradation which occurs in use; the US government has agreed to desalinate 4·546 million cubic metres per day (100 million gallons per day) before it passes to Mexico to permit further use.

World Health Organization, *International Standards for Drinking Water*, 1971.

Water supply. A complex subject best illustrated with the processes used to supply London with 'wholesome' water which implies that the proper PUBLIC HEALTH precautions have been taken. As London is supplied mainly by river abstraction from the Thames – 5000 million gallons per day in 1971 to a resident population of 6 million people, i.e. a consumption of 82·5 gallons (UK) per head per day (0·37 cubic metres per head per day) – and as this water must be purified, all typical water supply processes are covered in the example.

The presence of the coliform bacteria, *Escherichia coli*, is monitored in the source as an indicator of faecal pollution (◊ COLIFORM COUNT). In Figure 61 the count is 284 per cubic metre. The abstracted water is pumped to a reservoir for storage and settlement where faecal bacteria are reduced through natural processes and a large portion of the suspended solids settle out.

The water is taken from the storage reservoir via the draw-off arrangement *under* the dam base to an aeration basin so that it is fully oxygenated, to the primary filter and micro-strainers where most plankton and algae are removed. (Prior to this stage coagulation is often practised whereby colloidal particles which carry negative charges and thus will not naturally come together are coagulated by special agents and removed by sedimentation.)

Sand or other filtration follows and this is a most important stage as the sand is very much more than a fine strainer. The sand filter is a large shallow basin on which there is 1 metre of sand on a gravel base. The water is introduced at the top of the bed and drawn off at the bottom through collector pipes and channels. Filtration in this case takes place under gravity although pressure filters are also available.

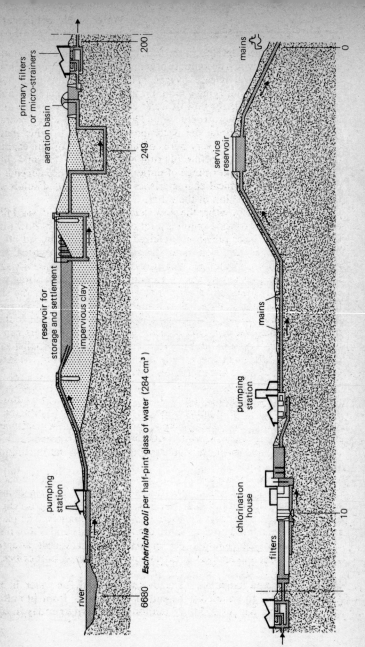

Figure 61. *Escherichia coli* count in London's water supply. (From Metropolitan Water Division, Thames Water Authority, *A Brief Description of the Undertaking*, 1972.)

Two separate layers develop in the bed: the top 2 millimetres comprise micro-organisms which decompose organic matter and consume nitrates and phosphates and release oxygen. This is the autotrophic layer which has a strong purifying action, on whose surface there is a muck film which can impede the flow of water considerably. Below the autotrophic layer is the heterotrophic zone (30 centimetres) where bacteria (non-pathogenic) continue the decomposition of the organic matter and render the water in a very pure condition. In the London example the *E. coli* count is 200 before filtering and 10 after. The sand filter, thus, performs a complex range of duties. Its performance depends on the physical and chemical characteristics of the suspended solids and the chemical composition of the water.

Filtration can be by either the slow sand filter or the rapid sand filter method. The slow method requires a lot of space and is not now often used as the dirt layer previously referred to must be skimmed off by draining the filter, and new sand spread on top every two or three months. The flow of water per unit area is low. The rapid gravity sand filter is 20 times faster and this is accomplished by backwashing with compressed air. The rapid filter is not nearly so efficient as the slow filter so coagulation is required as a pretreatment process. The rapid filter does not oxidize organic matter and nitrogen compounds so these may require removal at a later stage *or* anthracite sand filters can be used. These are a layer of large anthracite grains on top of smaller sand grains. These filters clog less rapidly as the muck adheres to the larger grains and thus larger filtration rates are possible.

The water with an *E. coli* count of 10 is now sterilized by chlorine injection at a rate proportional to the water flow. The *E. coli* count is now zero and the presumption is that the water is now wholesome. It is then pumped to the mains where it is distributed to the consumers. Note that service reservoirs and/or water towers are required to maintain adequate pressure in the supply mains and to iron out fluctuations in demand.

Another form of sterilization (not shown in Figure 61) is the use of OZONE, which is a powerful oxidizing agent and which also reduces colour, taste and odour in water. A dose as low as 0·0001 per cent or 1 part per million destroys all bacteria within 10 minutes. The ozone must be manufactured by electric-arc discharge on site and is thus more expensive than chlorination. No matter what method is adopted, extensive testing and analysis is carried out by all water authorities.

Water supply and irrigation. The overall demand for water in the UK is projected to double by the end of the century, from 14 million cubic metres per day in 1973 to 28 million cubic metres per day in 2001

(Water Resources Board). Rivers are used to supply domestic water and also receive effluent discharge from sewage, industry and agriculture. As demand for water increases, so more water must be made available to obtain water fit for consumption. This can be done by drawing water from reservoirs, by sinking WELLS, and by extracting water from AQUIFERS.

However, the two latter methods can have a serious effect on agriculture by lowering the water table. MODERN FARMING METHODS have already caused deterioration in the quality of soils (◊SOIL, FERTILITY AND EROSION OF). Large-scale irrigation will improve the situation, but some form of price regulation will probably be needed to curb the rise in domestic and industrial consumption and ensure that there is enough water for agricultural purposes.

Water Resources Board, *Water Resources in England and Wales*, vol. 1, HMSO, 1974.
M. Allaby, C. Blythe and C. Hines, *Losing Ground*, Earth Resources Research for Friends of the Earth, revised edn, 1975.

Watt. A measure of power or rate of energy supplied. It is defined as a joule per second ($J s^{-1}$). The energy can be in thermal form, in which case it is denoted by W_{th}. The distinction between thermal and electrical energy is important (◊ENERGY).

Wave power. The harnessing of the energy in the waves to generate electricity. Engineering assessments show that to get the equivalent amount of energy from 1 kilogramme of coal requires 1 tonne of sea-water to fall through 3 kilometres. Thus, the capital costs for the harnessing of wave power may be absurdly high.

Wells and boreholes. A hole drilled into the ground to tap an aquifer for water supplies. Shallow wells in soft formations can be dug by hand or with power augers. Deeper and/or resistant rock formations are drilled by machine. Once the well has been drilled it must be completed, i.e. the hole cased if necessary to prevent collapse with a slotted casing to allow the water to enter. The well is then developed by explosives, chemicals or compressed air to shatter the rock and increase the yield.

Once pumping commences, the safe yield is established. This can be restricted by salt-water infiltration in some areas. This means that a well's yield is not necessarily the maximum rate at which water can be extracted, but the maximum rate at which the quality of the water or the supply of other nearby wells is not affected.

Whale harvests. Whale fisheries serve as an archetypal model of over-exploitation. Figure 62 shows the situation. As the large whales were

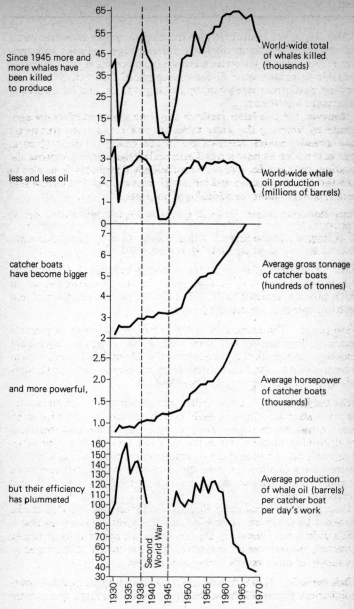

Since 1945 more and more whales have been killed to produce

World-wide total of whales killed (thousands)

less and less oil

World-wide whale oil production (millions of barrels)

catcher boats have become bigger

Average gross tonnage of catcher boats (hundreds of tonnes)

and more powerful,

Average horsepower of catcher boats (thousands)

but their efficiency has plummeted

Average production of whale oil (barrels) per catcher boat per day's work

Second World War

Figure 61 a and b. The process of extermination of whale species over the past forty-five years. (From P. A. Ehrlich and A. H. Ehrlich, *Population, Resources, Environment,* W. H. Freeman, 1970, p. 105.)

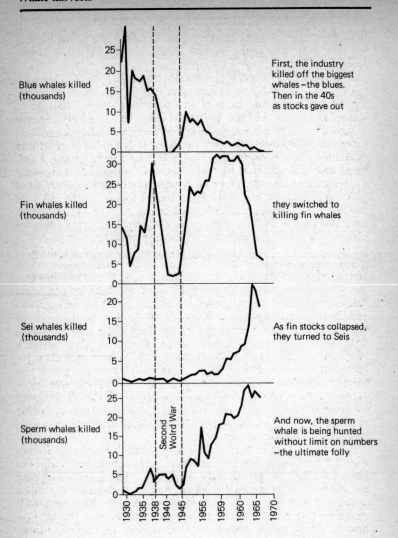

Blue whales killed (thousands)

First, the industry killed off the biggest whales – the blues. Then in the 40s as stocks gave out

Fin whales killed (thousands)

they switched to killing fin whales

Sei whales killed (thousands)

As fin stocks collapsed, they turned to Seis

Sperm whales killed (thousands)

Second Wolrd War

And now, the sperm whale is being hunted without limit on numbers – the ultimate folly hunted to near or total extinction, the industry shifted to harvesting not only the young of the larger species but also smaller species.

After the Second World War (indicated by the grey band in Figure 62) the International Whaling Commission (IWC) was established, with the brief of regulating harvests and protecting certain endangered species. In practice the IWC has turned out to be a relatively impotent organization with inadequate power of inspection and non-existent powers of enforcement. The IWC made an early fundamental error in establishing quotas on the basis of 'blue whale units' (bwu), a unit being one blue whale, or its equivalent (2 fin whales, $2\frac{1}{2}$ hump-back whales, or 6 sei whales). The total permitted quota was initially set at 16 000 units with the blue whale the preferred species. In the 1950s the number of blue whales caught declined sharply and the whalers turned their attention to the fin and hump-back whales. During the 1960s the blue whales and fin whale catch continued to decline, and in 1963 a Commission committee report warned that blues and hump-backs were in serious danger of extinction. It recommended that these species be totally protected and the fin whale catch limited. It also urged that the 'blue whale unit' be dropped in favour of limits on individual species. Predictably these warnings and recommendations were ignored and the total catch quota was limited to 10 000 bwu. By this time the fin whale harvest was estimated to be *three times* the MAXIMUM SUSTAINABLE YIELD. Subsequently the committee recommended a bwu total for 1964–5 of 4000; for 1965–6, 3000; and only 2000 for 1966–7, in order to allow recovery of the whale stocks. Again it was ignored. All four countries then engaged in Antarctic whaling – Japan, the Netherlands, Norway and Russia – voted against accepting the recommendation and continued to practise their trade according to the sound economic principle of obtaining a maximum return on invested capital (◊ECONOMICS).

By 1965, catches of blue and fin whales had declined even further and well over one-third of the estimated total population of seis was being harvested. By this time the Netherlands had given up whaling altogether and sold her fleet to Japan. Also in 1965 the IWC decided to limit the 1965–6 catch to 4500 bwu (way over the committee's recommendations) and tried, without success, to give the by now heavily fished sperm whales some protection. Subsequent further attempts to limit the take of fin, sei and sperm whales in Antarctic and other waters have failed. In 1966–7 the Antarctic whaling fleet catch was down to 3511 bwu (4 blues, 2893 fins and 12 893 seis). In 1967, a grand total of 52 046 whales were slaughtered, 25 911 of them sperm whales, and in addition Japan had killed 20 000 porpoises. In 1968

Norway was forced out of the whaling industry, leaving the field to Japan and Russia.

Setting aside any consideration of the aesthetics of such unrestricted slaughtering of these magnificent and intelligent animals, what can be said about the whaling industry's performance? For one thing, of course, their drive toward self-destruction tends to contradict the commonly held notion that people would change their behaviour if they realized that it was against their own self-interest. The whaling industry has operated against its own long-term self-interest continually since 1963, in full knowledge of what it was doing. Short-term self-interest, the lure of the 'quick buck', clearly is too strong to allow the long-range best interest of everyone to prevail. This is just one example of a cost–benefit analysis done over too short a term (P. A. and A. H. Ehrlich, *Population, Resources, Environment*, W. H. Freeman, 1970, p. 106).

(✥ANCHOVY FISHERIES.)

Wind. The movement of air parallel to the earth's surface caused by differences in atmospheric pressure and the earth's rotation. Above a certain height the wind blows at constant velocity in a direction parallel to the isobars – lines of constant pressure on the weather chart. The velocity is proportional to the isobar spacing – the pressure gradient – and is known as the gradient wind. This occurs above the gradient height, approximately 600 metres above ground.

Below the gradient height the wind is slowed down by ground effects, mainly friction, and is also stirred near ground level by obstructions so that eddies are set up which are very useful in diluting and dispersing air pollutants emitted at or near ground level.

Wood alcohol. ✥METHANOL.

World models. The attempts by Forrester and others to develop computer models of the world's economic and environmental condition for descriptive and (perhaps eventually) predictive purposes, as described in *The Limits to Growth* and *Mankind at the Turning Point*, have caused a storm of outraged reaction from social scientists in general and economists in particular. The assertions of the anti-Forester camp can be summarized as follows:

1. '. . . the model neglects to examine and use the relevant theory and empirical evidence generated by the discipline of economics, and . . . as a consequence it also neglects technical progress' (Pavitt, 1973).

2. Others (Ricardo, Malthus) have concluded that growth could not continue indefinitely because of physical limits and they have turned out to be wrong.

3. Arguments that growth in the industrialized countries should stop are unwarranted; '. . . we should also consider the possibility that we should go on working, not to increase our own consumption, but to better the lot of the poor majority of the world' (Pavitt, 1973).

4. Assumptions made in formulating the models are gross over-simplifications, and in particular the aggregation (in *The Limits to Growth*) is hopelessly unrealistic.

Of these objections, only (4) has any real substance. The rest are petulant (1), silly (2) and almost incredibly naive (3).

While it is undoubtedly true that practically all the assumptions made in the formulation of the computer world models produced so far are gross ones, and that undoubtedly some of them are invalid, one should be clear that this is not the basic reason for the storm of protest that the work has produced. The reason is that while the assumptions made by the computer modellers are in fact no more gross or unrealistic than those assumed by economists, they are, unfortunately, *different* assumptions.

Models of technological innovation, substitution and mineral discovery that are used in economics bear roughly the same order of resemblance to reality that 'economic man' bears to a real person. Their substantiation rests on no more than an argument of the sort 'It stands to reason that . . .'. Such models have assumed axiomatic status in economics and any work which questions or ignores them is anathema to the traditional economist.

Nevertheless, because the majority of criticism directed at world-modelling exercises is misdirected does not mean that there are not serious defects in both the methodology and the models in question. In particular there are indications that the authors, like the economists who are so opposed to them, are in danger of confusing their models with the real world. At the present stage of the work this would be most foolish. Indeed it is very doubtful whether truly predictive models of world behaviour could ever be developed. In the meantime it is best to view both traditional economic theory *and* world models as useful parables from which some insight as to possible future options may, with care, be gleaned. Perhaps the true value of world models is that they call into question the implicit assumptions of traditional economics and that from this conflict some progress may ensue. It is probably only fair to let Professor Forrester have the last word:

A lot of people believe this [building of models] is impossible, because they feel that one cannot make a model of anything as complicated as a social system, and of course they are partially correct – a model is a great simplification. But we have no choice about the use of models. The mental image that we use for passing laws, for running cities, for operating a government – these mental images are models – because one does not have a city, or a country, or a world in one's head, one only has certain images – which are models – and so the question now is: Is the model adequate?, is it the *best* model which we can make?

– and the answer is no! it is *not* the best model which we can make! (*Interview from Open University Technology Foundation Course (T100) television programme on 'Limits to Growth'.*)

D. H. Meadows, D. L. Meadows, J. Randers and W. W. Behrens, *The Limits to Growth*, Earth Island; Pan, 1972.
M. Mesarovic and E. Pestel, *Mankind at the Turning Point*, Hutchinson, 1975.
K. Pavitt, 'Some consequences of physics – based research', *Physics Bulletin*, January 1973.

X

X-rays. Ionizing or electromagnetic radiations of the same type as light but with much shorter wavelength. The absorption of the rays depends on the density of the material; as bone is denser than flesh this allows x-ray photographs to be taken. All radiation exposure carries risks (◊IONIZING RADIATION). However, medical irradiation is voluntary in the sense that the benefits incurred by the diagnostic x-rays far outweigh the risks. Having said that, pregnant women and/or fertile women are not normally exposed to x-rays as the foetus is particularly susceptible in the first instance and the germ cells in the second.

Y

Yeasts. Unicellular fungi which are able to multiply asexually by a budding process. Yeasts are of great importance in FERMENTATION, the manufacture of pharmaceuticals and SINGLE CELL PROTEIN. One major strain used for single cell protein from molasses and the sugar in the spent liquors from sulphite pulp manufacture is *Candida utilis*, which has an approximate 50–60 per cent protein content.

Guide to Further Reading

General

The following books provide a useful general introduction to the study of the environment. The last two books on the list present a view diametrically opposed to that of most environmentalists.

Fairfield Osborn, *Our Plundered Planet*, Faber, 1949.

William Vogt, *Road to Survival*, Gollancz, 1949.

Kenneth E. Barlow, *The Discipline of Peace*, 2nd edn, Charles Knight, 1971.

Rachel Carson, *The Silent Spring*, Hamish Hamilton, 1963; Penguin Books, 1970.

Barbara Ward and Rene Dubos (eds.), *Only One Earth*, André Deutsch; Penguin Books, 1972.

P. R. Ehrlich and A. H. Ehrlich, *Population, Resources, Environment*, W. H. Freeman, 1970.

Barry Commoner, *The Closing Circle*, Cape, 1972.

Royal Commission on Environmental Pollution, *Reports*, HMSO, vols. 1–3, 1972; vol. 4, 1974; vol. 5, 1976.

Brian Harvey and John Hallett, *Society and Environment: An Introductory Analysis*, Macmillan, 1976.

Wilfred Beckerman, *In Defence of Economic Growth*, Jonathan Cape, 1974.

John Maddox, *The Doomsday Syndrome*, Macmillan, 1972.

Economics and the Environment

Charles Carter, *Wealth: An Essay on the Purposes of Economics*, Penguin Books, 1971.

E. F. Schumacher, *Small is Beautiful*, Blond & Briggs, 1973.

J. K. Galbraith, *The Affluent Society*, Hamish Hamilton, 1958; Penguin Books, 1970.

J. K. Galbraith, *The New Industrial State*, André Deutsch, 1972; Penguin Books, 1968.

J. K. Galbraith, *Economics and the Public Purpose*, André Deutsch, 1974.

E. J. Mishan, *The Costs of Economic Growth*, Staples Press, 1967; Penguin Books, 1969.

Politics and the Environment

Geoffrey Vickers, *Freedom in a Rocking Boat*, Allen Lane: The Penguin Press, 1970; Penguin Books, 1972.

Robert L. Heilbroner, *An Inquiry into the Human Prospect*, Calder & Boyars, 1975.

Alvin Toffler, *Future Shock*, Bodley Head, 1970; Pan, 1973.

Jeremy Bugler, *Polluting Britain*, Penguin Books, 1972.

The Use of Computer Models

D. H. Meadows, D. L. Meadows, J. Randers and W. W. Behrens, *The Limits ot Growth*, Earth Island; Pan, 1972.

J. W. Forrester, *World Dynamics*, Wright Allen, 1971; Wiley, 1972.

M. D. Mesarovic and E. Pestel, *Mankind at the Turning Point*, Hutchinson, 1975.

World Models – Sense or Nonsense, Open University, 1974.

Technology: A Third Level Course, System Modelling Unit 16, Open University, 19

Energy

P. Chapman, *Fuel's Paradise*, Penguin Books, 1975.

US Energy Prospects – An Engineering Viewpoint, US National Academy of Engineering, 1975.

W. C. Patterson, *Nuclear Power*, Penguin Books, 1976.

A. B. Lovins, *World Energy Strategies*, Earth Resources Research for Friends of the Earth, 1973.

A. B. Lovins, *Nuclear Power: Technical Bases for Ethical Concern*, Friends of the Earth, 1975.

John Maddox, *Beyond the Energy Crisis*, Hutchinson, 1975.

A Critique of the Electricity Industry, Energy Research Group Research Report ERG 013, Open University, March 1976.

Andrew Porteus, *Saline Water Distillation Processes*, Longman, 1975.

Agriculture and Food

Gerald Leach, *Energy and Food Production*, International Institute for Environment and Development, 19

Public Health

Environmental Control and Public Health, Course PT 272 (Chairman and General Editor: A. Porteus), Open University, 1973.

Anthony Tucker, *The Toxic Metals*, Pan; Ballantine, 1972.

Health Hazards of the Human Environment, World Health Organization, 19 .

G. L. Waldbott, *Health Effects of Environmental Pollutants*, C. V. Mosby, St Louis, 1973.

Colin Walker, *Environmental Pollution by Chemicals*, Hutchinson, 1971.

Andrew Porteus, Chris Pollitt, Keith Attenborough, *Pollution – The Professionals and the Public*, Open University, 1976.

Organizations concerned with conservation

Alternative Society
9 Morton Avenue
Kidlington
Oxford
Kidlington 3413

Association for the Preservation
of Rural Scotland
1 Thistle Court
Edinburgh 2
031–225 6744

British Society for Social
Responsibility in Science
9 Poland Street
London W1V 3DG
01–437 2728

British Trust for Ornithology
Beech Grove
Tring
Herts.
044–282 3461

Centre for Environmental Studies
62 Chandos Place
London WC2
01–240 3424

Committee for Environmental
Conservation (COEnCO)
29–31 Greville Street
London EC1N 8AX
01–242 9647

Committee for Environmental
Information
438 N Skinker Boulevard
St Louis
Missouri 63130
USA

Conservation Society
12 London Street
Chertsey
Surrey KT16 8AA
Chertsey 60975

Consumers Association
14 Buckingham Street
London WC2
01–839 1222

Council for Nature (Wildlife)
Zoological Gardens
Regent's Park
London NW1 4RY
01–722 7111

Council for the Protection of

Rural England
4 Hobart Place
London SW1W 0HY
01-235 4771

Council for the Protection of
Rural Wales
Mr Simon Mead
Meifod
Powys
Meifod 383

Countryside Commission
John Dower House
Crescent Place
Cheltenham
Glos. GL50 3RA
0242 21381

Environmental Consortium
10-11 Great Newport Street
London WC2
01-836 0908

Friends of the Earth (FOE)
9 Poland Street
London W1V 3DG
01-434 1684

Henry Doubleday Research
Association
Convent Lane
Bocking
Braintree
Essex
Braintree 1483

Intermediate Technology
Development Group
Parnell House
25 Wilton Road
London SW1V 1JS
01-828 5791

International Planned
Parenthood Federation (IPPF)
18 Regent Street
London SW1Y 4PW
01-839 2911

International Union for the
Conservation of Nature
1110 Morges
Switzerland
(021) 714401

National Society for Clean Air
136 North Street
Brighton
East Sussex BN1 1RG
0273 26313

National Trust
42 Queen Anne's Gate
London SW1
01-930 0211/1841
Nature Conservancy Council
19-20 Belgrave Square
London SW1X 8PY
01-235 3241

Noise Abatement Society
6 Old Bond Street
London W1
01-493 5877

Population Countdown
Elsley House
24-30 Great Titchfield Street
London W1A 2LB
01-580 7331

Population Stabilisation
5 Riverdale Road
East Twickenham
Middlesex
01-892 0238

Royal Commission on
Environmental Pollution
Church House
Great Smith Street
London SW1
01–212 8620

Royal Society for the Protection
of Birds
The Lodge
Sandy
Beds.
0767 80551

Society for the Promotion of
Nature Reserves
The Green
Nettleham
Lincolnshire ON2 2NR
0522 52326

Soil Association
Walnut Tree Manor
Haughley
Stowmarket
Suffolk IP14 4RS
044–970 234

Third World First
PO Box 59
4 Marston Ferry Road
Oxford
Oxford 54006

United Nations Association
93 Albert Embankment
London SE1 7TX
01–735 0181
(Has nearly 400 local groups)

UNICEF
99 Dean Street
London W1V 6QW
01–734 2398

Vegetarian Society of the
UK Ltd
53 Marloes Road
London W8 6LD
01–937 7739

Voluntary Committee on
Overseas Aid and Development
(VOCAD)
Parnell House
Wilton Road
London SW1V 1JS
01–828 7611

World Development Movement
Bedford Chambers
Covent Garden
London WC2E 8HA
01–836 3672

World Wildlife Fund (British
Appeal)
29 Greville Street
London EC
01–404 56911

Magazines Concerned with Environmental Problems

Ecologist
73 Molesworth Road
Wadebridge
Cornwall PL27 7DS

Environment
Publication Department
17 Ridgmont Road
Bramhall
Cheshire

Other Books in the Series

Arrow dictionaries have been written and designed to open up expanding specialist areas of knowledge to the non-specialists, both students and general readers.

Other books currently in the series are:

Professor David Dineley, Professor Donald Hawkes, Dr Paul Hancock and Dr Brian Williams
EARTH RESOURCES
a dictionary of terms and concepts

Tim Congdon and Douglas McWilliams
BASIC ECONOMICS
a dictionary of terms, concepts and ideas

Dr Anthony Hyman
COMPUTING
a dictionary of terms, concepts and ideas

Professor David Dineley, Professor Donald Hawkes, Dr Paul Hancock and Dr Brian Williams

EARTH RESOURCES

a dictionary of terms and concepts

A full understanding of the natural commodities on which industrial society depends has never been more important than it is today.

This dictionary provides an introduction to the terms used in the study of the world's resources – from metallic ores and chemical deposits to the hidden wealth of oil and gas and to our one great re-usable resource, water.

For the student who needs a ground knowledge of the various branches of the subject – mineralogy, geology, and oceanography – and for the amateur interested in the origin, character and technology of the world's natural wealth, *Earth Resources* will be an invaluable handbook.

Professor David Dineley is Head of the Department of Geology at Bristol University.

Professor Donald Hawkes is Head of the Department of Geological Sciences at the University of Aston, Birmingham.

Dr Paul Hancock and Dr Brian Williams are Lecturers in Geology at Bristol University.

Tim Congdon and Douglas McWilliams

BASIC ECONOMICS

a dictionary of terms, concepts and ideas

As never before, the world of economics is headline news, influencing our political, professional and personal lives.

As its importance grows, so do its complexities. This dictionary aims to provide both the student and the general reader with a working knowledge of the basic terms and concepts of economics and finance.

The emphasis throughout is on comprehensibility and clarity. With an easy-to-use system of cross-referencing, the dictionary will be indispensable not only as a work of reference but also as a lively introduction to the study of economics.

Tim Congdon is economics correspondent of The Times *and winner of the Wincott Press Award for the Young Financial Journalist of the Year in 1976.*

Douglas McWilliams is an economic forecaster for the Confederation of British Industry.

Dr Anthony Hyman

COMPUTING

a dictionary of terms, concepts and ideas

The working of society today depends increasingly on the computer but how many of us really understand this machine that dominates our lives?

This dictionary is designed specifically to help the student and layman come to terms with the computer in all its forms. Anthony Hyman defines terms, explains concepts and explores ideas – all in a language that makes the technical accessible to the non-technical reader.

The prime aim of the book is to equip the reader with a working knowledge of computer functions, systems and applications so that he may relate his knowledge to practical day-to-day use.

Dr Anthony Hyman has worked on computers, computing and related topics for 25 years. He holds 50 patents and has published scientific papers on many aspects of the subject.